THE POSTMISTRESS

OF

SADDLESTRING, WYOMING

Edgar M. Morsman, Jr.

Edgar M. Morsman, Jr., Publications
Deephaven, Minnesota

Additional copies of this publication and order information can be obtained by contacting Edgar M. Morsman, Jr., Publications, 4011 East Valley Road, Deephaven, Minnesota 55391-3667; phone: 612-476-4919; fax: 612-476-9031.

*To Beth, Jim, Jeff, and Tim who helped me live a great
deal of what follows.*

CONTENTS

PREFACE

What is it that defines the unforgettable character, the one person in thousands who will be remembered with a mixture of joy, admiration, and fondness by almost everyone he or she meets? Certainly, such rare personalities inspire both fascination and, in some cases, awe. Many in their orbit crave their respect, if not their affection.

Such rare people seem to achieve their potential, giving them a fullness of character and substance. Yet they are not necessarily heroic or saintly. They seem to have a natural curiosity that makes them interested in many facets of life. Above all, they are always themselves interesting.

Perhaps they are interesting because they tend to live each day to the maximum, to extract the potential from the time allotted. John Dryden may have had such people in mind when he wrote:

> Happy the man, and happy he alone,
> He who can call today his own;
> He who, secure within, can say,
> Tomorrow, do thy worst, for I have lived today.

These rare individuals seem to be able to shrug off today's disaster and then anticipate what the next day might bring. Perhaps they have achieved balance, inner peace, or, maybe, just cheerful resignation. They seem to accept life's vicissitudes with grace and a sense of humor. Or maybe what defines them is that someone feels a need to preserve their memories for posterity.

Henrietta "Hank" Horton was such a person, and this book is about her and some of the people that surrounded her—Jack and Trudy Horton and Dean Thomas. Although each was highly individualistic, their frequent interactions also defined them as did the place that they all called home for at least part of their lives: the HF Bar Guest Ranch. There were many others who were part of Hank's life, of course, and a few will be mentioned. I apologize for the many obvious omissions.

This book is not a biography, for much that is important is missing. Although all four traveled extensively or spent significant parts of their lives outside of the state, I think they always retained a large piece of Wyoming in their characters. Consequently, most of the stories take place in Wyoming, so considerable biographical detail will be missing.

This book does not delve into the psychology of the characters. I have too much respect for these individuals to attempt such an intrusion. Suffice it to say they were complex people who, like most interesting people, were not easy to read.

They were not perfect. Any strong willed person can be difficult at times, and most people, including me, knew them only in a vacation setting, which can be distorting.

They were not invincible. They had the normal share of failure, and three of the four died prematurely from cancer.

But they were never boring. People delighted in their presence and considered time spent with them special.

If One Martini Is Good. . .

However, even given the uniqueness of these four, what could possibly cause a retired banker who has written books only on commercial lending to attempt to capture the spirit of such incredible people? There is only one logical answer: Beefeater martinis and cabernet sauvignon. In early July 1996, my wife, Beth, and I were enjoying our accustomed two-week vacation at the HF Bar Guest Ranch near Buffalo, Wyoming. That most engaging time of day, the cocktail hour, had been both prolific and soothing. As one of my former bosses once said, one martini is just right, two are too many, and three are not enough. I don't recall where I was on that continuum.

Dinner had been al fresco with copious cabernet, and I was in a felicitous mood aided by having forgotten to sign up for the evening ride. Most guests come to the ranch for the horseback riding, and although I have ridden since the age of five, I have always considered swinging onto the back of twelve hundred pounds of stupidity and unpredictability a rather extreme form of entertainment, particularly when I must pay for the pleasure. I can cover the same ground on foot, enjoying the birds, flowers, and panoramas without the tedious codependency of a large equine companion. Fighting words in Wyoming—I live in Minnesota.

The air was cool and dry with that deep azure sky filled with billowing cumulus clouds that are the hallmark of a Wyoming evening. I was chatting amicably with Peter Constable whose extended family covers four generations of ranch goers. Naturally, the conversation eventually turned to, "Remember the time when Hank and Dean...."

Someone said we should write those stories down before they were all forgotten. Have you ever heard that warning beep that heavy equipment makes when it is backing up? I heard the same sound but lost it amidst the general bonhomie of the moment. The next thing I knew, I had been volunteered. Peter graciously offered to help in the collection of the stories.

Margi Schroth, the current owner of the HF Bar, was passing by our table, so we flagged her down and enlisted her support and encouragement. She wisely noted that we had better get on with it and fast. Being in a somewhat lethargic mood, I asked what was the rush and she pointed out that the people who knew Hank, Dean, Trudy, and Jack are an endangered species.

Hank would be in her mid nineties today and Dean in his late eighties. People that knew them well, like George and Freddie White, my uncle Truman and aunt Sis Morsman, my aunt's brother, Tom Baird, to name just a few, are long gone. They not only knew the stories. They were part of many of them.

Consequently, this small collection of stories is in no way complete. Many of you whom I have never had the privilege of meeting, particularly those in the August crowd, will undoubtedly spot glaring omissions. Nevertheless, I hope there is enough to give substance to these characters.

Contributors

I am deeply indebted to Peter Constable for his interviewing diligence in extracting stories from people inaccessible to me. A partial list of contributors follows. I regret any omissions and, perhaps more important, people who could have contributed but were not asked through ignorance. I believe the geographical distribution of the contributors connected by the small locus of Saddlestring, Wyoming, is a tribute to the magnetism of the characters in the book.

David and Julie Allen	Gaithersburg, Maryland
Peter and Elinor Constable	Washington, D.C.
Philip and Melinda Constable	Boston, Massachusetts
Robert and Jennifer Constable	Boulder, Colorado
Anne Dixon	Longmont, Colorado
Pam and Mike Dixon	San Rafael, California
Dick and Dottie Deuser	Minocqua, Wisconsin
Bronco and Botsy Jones	New York, New York
Mac and Jane McDougal	Peoria, Illinois
Susanna Taylor Meyer	Woodbury, Minnesota
John and Julie Schroeder	Waterloo, Nebraska
Mary Schroeder	Lyme, Connecticut
Margi Schroth	Saddlestring, Wyoming
Shirley Taylor	Kirby, Montana
Kelly and Mary Welch	Tucson, Arizona
Flora Westin	New York, New York
George and Jo White, Jr.	Portola, California

Any Story Worth Telling Is Worth Enhancing

Much time has passed since these events occurred, and in some cases, these stories have been recounted through several generations. Because I cannot personally vouch for the accuracy of every event, I have omitted, in most cases, the names of specific participants to protect both the innocent and the guilty. Having known the participants, I do feel, nevertheless, that the stories are directionally correct.

On a final note, the title of this book was suggested by Tom Baird, an old friend of Hank's, who dedicated one of his novels to the postmistress of Saddlestring, Wyoming.

Edgar M. Morsman, Jr.
June 1998

⟨THE EARLY YEARS⟩

I f you were to drive west on Interstate 90 from the Minnesota–South Dakota border to the Missouri River at Chamberlain, you would first be impressed with the flatness of the land. This impression would soon give way to boredom, then tedium, and finally desperation if you were at all fatigued.

Although planted in corn and soy beans, the land is nowhere near as fertile as the land of Iowa or Illinois. Texts in the early 1800s referred to the area between the Missouri River and the Rocky Mountains as the Great American Desert. In that era, pioneers would have been on horseback trudging through lush blue stem and Indian grass knee to waist high.

Today's travelers can barely imagine the tall grass prairie. Multihued tall grasses undulating in the breeze would stretch as far as the eye could see. The subtle variety of green colors waving under endless blue skies with massive cumulous clouds would match the vastness of the oceans. And the silence would be profound, broken only by the cry of a raptor and the rustling of grasses. Wallace Stegner wrote of "the grassy, green, exciting wind, with the smell of distance in it."

As you cross the Missouri River, you may notice a subtle change in the land ahead. Rolling hills intersected by draws that drain the occasionally excessive rainfall back into the Missouri replace the dull flatness. The fields of corn and beans give way to wheat and hay. The landscape becomes dotted with cattle as farming succeeds to ranching.

Had you traveled this route in the early 1800s, you would have experienced a gradual change in the prairie grasses between the Big Sioux River where Sioux Falls is now located and the Missouri River at Chamberlain. The tall prairie grass would slowly surrender to short grass plains dominated by buffalo grass and blue grama that would sweep to the Rocky Mountains.

As your car climbs out of the Missouri valley heading west, you pass a statue of a buffalo at Al's Oasis, a rather pathetic reminder of the immense herds of bison that once covered the plains. Other mammals were once abundant that we do not today think of as plains animals. In their journals, Lewis and Clark describe the herds of elk that sustained them and the occasional and formidable encounter with the grizzly bear. With the loss of habitat, these animals have retreated to mountain refuges.

I have always felt that the West begins at the Missouri River with its subtle change in land configuration and aridity that once announced the change from tall grass prairies to short grass plains. Civilization has wrought a wrenching change in the ecology and culture of the region, transforming it in the space of a mere one hundred years, a wink in geological terms, over the mid-nineteenth to the mid-twentieth century. Anyone living in the area during this period truly had a foot in two different worlds.

As you continue west on I-90, the Badlands near Wall provide some relief from the rolling hills, but a detour of several hours is required to gain any sense of their unique beauty. Finally, that outpost of the Rocky Mountains, the Black Hills, rises from the horizon.

At an elevation of 7,242 feet, Harney Peak in the Black Hills is the highest point between the Rockies and the Pyrenees. Yet it is a mere foothill compared with the mountains further west. The interstate highway skirts the Black Hills to the north, enters Wyoming, and leaves the Black Hills formations near Sundance.

When you leave the Belle Fourche (inexplicably pronounced "Foosh" in the area) drainage near Moorcroft, you enter the Powder River Basin where the waters run north to the Yellowstone that, in turn, meets the Missouri on the Montana–North Dakota border. As you continue west under an unrelenting sun, you would probably agree with a book on my desk, *Modern Geography*, published in 1830 that describes the land as a "vast, elevated and barren waste, destitute of timber and vegetation."

Barren waste is probably a little harsh, but a little bit does go a long way. Rolling grasslands appear pale green to yellow depending on how much rain falls in late spring and early summer. Distinctive buttes lend a certain wildness to the character of the land. The grass in the bottom of the draws is darker, tracing the drainage and displaying how precious the sporadic rainfall is.

Near Gillette, exposed rock adds to the wildness of the terrain. The short grasses show subtle variations in color as clumps of sage add a bluish green, while clover produces patches of emerald green. Shimmering copses of cottonwoods delineate the main water routes such as the Powder River, a mile wide and an inch deep, and creeks with names such as Wild Horse, Dead Horse, Indian, Crazy Woman, and Dry Creek.

You will see sparse herds of cattle, and the occasional antelope will gaze at your car with dull curiosity. Man-made signs of life relate primarily to vast beds of coal formed in the Cretaceous or Tertiary periods, just yesterday in geological terms. Wyoming coal is strip mined and transported to the Midwest by rail, although some provides power on site. A few oil pumps dot the landscape looking like giant praying mantises with heads bobbing over their victims. This wealth of energy inevitably brings some ugliness—power lines, coal trains, trailer courts, and junk yards.

The heat, the endless short grass plains, and the unsightliness may make you wonder if you have made a horrible mistake in undertaking this trip. A little bit west of Gillette, as you top a rise, your faith is restored by the first sight of the Big Horn Mountain range, a vision so breathtaking, so in contrast to the plains, that you can't help but think you are entering a different world.

The Big Horn Mountains are approximately 120 miles long and 30 to 50 miles wide, a fairly small mountain range. Although the alpine region is even smaller, the upward thrust of its peaks is so impressive, the mountains seem bigger than they really are. Cloud Peak reaches 13,166 feet, the second highest in Wyoming. When you top the rise west of Gillette and get your first, unforgettable view, you are gazing directly at these granite peaks, which even into early summer normally have a dazzling cover of snow.

The oldest rocks in the core of the mountains were formed several billion years ago. Ancient seas flooded the area and then subsided leaving layers of mineral and organic sediment over the basin. Like the rest of the Rockies, the Big Horns were formed by a slow uplifting caused by the Continental plate drifting westward, colliding with and overriding the Pacific plate. This process began a mere eighty million years ago.

The Big Horns are very young mountains in geological terms, which accounts for some of their beauty. As the mountains were thrust upward by the drifting plates, the erosive action of wind and water fought to grind them down. A geologist named Darton, for whom a peak is named, estimated that Cloud Peak once reached twenty-five thousand feet but has lost seven thousand feet to the inexorable grinding of wind, ice, and water. Erosion carried away the softer sediments spreading them in the basins and exposed the inner granite core that the traveler finds so awe inspiring.

The most dramatic erosion occurred six to nine thousand years ago during the last Ice Age. As the climate cooled, snow accumulated and then turned to ice. Ice began to flow down the stream courses creating U-shaped bottoms. These glaciers deposited crushed rock as they moved along and left piles of rock where they stopped, producing the lateral and terminal moraines visible today.

In the high mountains, snow and ice accumulated against granite walls, froze, and then pulled away leaving gouged-out scoops called glacial cirques. If this action extended up hill, over time it would leave a series of lakes.

As the glaciers receded, life crept in to reclaim the land, tentatively at first, forming bogs, then meadows, and finally forests as the land dried out. Today you can see a mountain range crowned by a classical alpine region with vertical granite peaks and numerous sculpted cirques. Slabs of foothills rise from the plains followed by more rolling foothills. Great canyons gash through the hills formed by streams that drain the snow, ice, and rain from the peaks.

The original inhabitants of this land are now known as the Sioux, Crow, and Cheyenne. Their history is obscure. Near a pass in the northern part of the range is a man-made formation called the Medicine Wheel. It is a circle, seventy feet in diameter, with twenty-eight spokes radiating from a central hub. In the early 1970s, an astronomer from the University of Colorado, John Eddy, discovered that certain rocks were aligned with the sun at the solstice. The formation is so old, it is not mentioned in the histories of Native Americans.

The Sioux, Crow, and Cheyenne were a nomadic stone-age culture when western civilization arrived. As you enjoy your first glimpses of the Big Horns, you can easily understand how readily a well-watered region of grassy valleys and pine forests supported the game necessary for a nomad's survival.

Then, a steady migration of fur trappers, explorers, traders, prospectors, and settlers began, and finally an army was sent to protect them. With the natural food supply disrupted and threatened, clashes were inevitable.

The major clashes in the area were brief, yet pivotal. In 1866, Red Cloud amassed an overwhelming force and annihilated Brevet Lieutenant Colonel William J. Fetterman and his eighty-man detachment as they were lured too far from Fort Phil Kearny. Although the nation was shocked, it was too exhausted from the Civil War to retaliate, and the government abandoned the fort leaving the native tribes in an uneasy peace for ten years.

As the nation recovered from the war and western expansion continued, the military once again followed the migration, and serious conflict loomed. Proving that strategies hadn't changed in the military in thousands of years, another foolhardy officer named George Armstrong Custer lead his men to annihilation on the Little Big Horn in 1876.

As triumphant as these victories were for the Native Americans, they also hastened their end. The nation was once again outraged, but this time, the government had the time, the will, and the wherewithal to respond.

The Native Americans achieved these victories through an incredible political leadership that temporarily combined the warfaring capabilities of tribes and bands that normally had nothing to do with each other. Once these factions were amassed, they needed the leadership to strike fast and overcome the enemy. But as a nomadic people living off the land, they could not assemble many people in one area for a sustained period of time. After their victories, the tribes dispersed.

That they achieved such victories at all is testimony to their military skills. Operating at the peak of perfection both politically and militarily is not the norm of any people, regardless of their prowess. Inevitably, these great tribes were overcome by repeating rifles, a quartermaster corps that could sustain an army in the field indefinitely, and starvation.

With the Native American threat neutralized, farmers and ranchers moved in, and disputes over water and grazing rights replaced clashes with native peoples. These disputes escalated to the Johnson County Cattle War of 1892. As more people settled in the area, lumbering began putting more pressure on the resources of the Big Horns. The Cloud Peak reservoir was constructed in the 1880s and other water diversion projects followed, which were grand engineering and political achievements.

Anyone living in or visiting north central Wyoming cannot help but be affected by the Big Horns. The beauty has drawn countless artists. Yet many conflicts have occurred and many lives lost because of its limited resources. These mountains affect people in strange and different ways, but no one who has lived in their shadow or traveled through their mysteries can say they have not been affected and, perhaps, even changed.

Frank "Skipper" Horton may have traveled the same route as I-90 follows today. However, Skipper would have been traveling at the turn of the twentieth century. Unique places seem to attract unusual people and Skipper was no exception. He was a doctor who had dreamed of becoming a surgeon, but colorblindness kept him out of the operating room.

Skipper was not a tall man, yet his intensity, boundless energy, and penetrating eyes made him seem bigger. When talking to you, he made you feel that you were the sole focus of his interests and thus could be incredibly persuasive, a talent that served him well as a congressman from Wyoming.

Skipper spent some time reconnoitering the area and ultimately bought land thirteen miles west of Buffalo from a widow whose husband had been killed in a sawmill accident. The best place to view the property is from the top of a crenellated and turreted sandstone formation named Castle Rock that connects the plains to the foothills via a series of hills ending in a dramatic saddle formation.

Skipper would not have been the first person to find a vantage point on Castle Rock. At its highest point, you can see 360 degrees surveying both mountains and plains. The ground is littered with flint chippings where ancient arrow makers practiced their craft on a spot that provided both security and intense beauty.

As you gaze westward toward the mountains, you notice two deep gashes in the slablike foothills from which the north and south fork of Rock Creek flow and ultimately join on the ranch property. The south fork rises from bogs under the Ant Hill, a mountain feature in the alpine region. The north fork rises naturally from drainage in the foothills but is enhanced by flows diverted from the watershed to the north, a rather incredible construction and political feat.

The immediate foothills are slablike features tilting toward the alpine region covered by the dark green of ponderosa and lodgepole pine. Erosion has exposed jagged rock varying in color from salmon to light gray. A narrow intrusion called the Speedway thrusts its light green short grass prairie form into the foothills, but it quickly surrenders to the coniferous forest. South of the Speedway is a deep canyon of red clay and salmon-colored granite called, as you would expect, Red Canyon.

Behind the first slabs of foothills are pine-covered hills rising to meet the granite peaks of the alpine region. The colors are green, gray, and black depending on the sun and cloud cover. With its sharply defined cirques and rivulets of snow disappearing into them, the alpine region is breathtaking.

To the south, grass-covered hills with rich red soil drift into the foothills. Hidden in these hills is the residue of an ancient limestone tower now called the Pillar of Salt. A woman of formidable intelligence and curiosity once took five rocks from the ground and had them analyzed by the head geology curator at the Smithsonian. One, a piece of limestone was formed under water. Two concretions of sandstone and phosphate were formed underground. A piece of basalt and tufa were both formed by volcanoes, one from a slowly cooling lava flow and the other from being blown into the air by a volcanic explosion. All were formed in different eras and levels but were all found lying on the ground together. A geologist's dream.

To the east, the plains are light green with slashes of red exposed earth, blue waters of Lake de Smet, and dark green cottonwood and aspen along water courses. Farming occasionally produces fields of emerald-green alfalfa. Running to the horizon, the plains are a rolling series of hills looking much like a vast ocean. The rising or setting sun accentuates the hills with pockets of shadows that emphasize the natural relief. To the south, the gigantic form of Pumpkin Butte rises from the plains, a natural landmark.

As your focus narrows from the broad vistas of mountains and plains, you will notice an impressionistic carpeting of flowers and vegetation. In late spring and early summer, it is as if the Maker had used a palette of lupine, flax, beards tongue, harebell, and penstemon for blues; yarrow, nodding onion, bed straw, moss campion, and mariposa lily for whites; and arrow leaf, balsam root, stone crop, prince's plume, and wallflower for yellow.

Castle Rock is the kind of place where you can cleanse your mind and renew your spirit. However, the achievement of tranquillity and serenity takes discipline. First, you must scan the vistas and allow their beauty to push most of the reality of the day from your consciousness. Then, you must listen to the silence broken only by the rustling of the wind through pine and short grasses and the occasional cry of a Clark's nutcracker or prairie falcon. You then experience the touch of the earth and the movement of air currents with different temperatures. The currents bring occasional and subtle fragrances of blooming lupine and of sun on pine. If you are disciplined enough to give up your mind to sight, touch, sound, and smell simultaneously, you will find yourself at total peace and in harmony with nature.

Unfortunately, the ambiance is not always so benign. At fifty-five hundred feet in altitude, you are subject to typical mountain weather patterns. A beautiful day can become a violent thunderstorm in hours. Days on end can be searingly hot turning the grass to a light brown, or the entire state can be covered with dull gray clouds with a drizzling coldness that invades your body to the core. And the pleasant season is short. Some wag once described the mountains as ten months of winter and two months of house guests, an accurate statement in Skipper's case. Winters can bring crystal clear skies with bone chilling cold or dull gray, dispiriting ugliness. Blizzards can deposit impassable snows and Chinook winds will melt them off turning the ground into boot sucking mud.

Skipper was a gregarious, extroverted man who soon drew a number of house guests. Finding the land so enchanting, the guests built cabins along the north fork of Rock Creek. When his guest total reached twenty-six, Skipper concluded that dudes might be a safer business than cattle, so he incorporated the HF Bar Guest Ranch in 1902, the second oldest in Wyoming.

Skipper had his feet planted in two worlds. When Skipper arrived in Wyoming, Custer's defeat was just twenty-five years ago and the Johnson County Cattle War had happened a mere eight years earlier. The men in the corral and the packers who took guests into the mountains had been a part of range wars or knew the participants. A little bourbon would bring forth a steady stream of stories, although the storytellers were fairly silent about which side they positioned themselves or where they were when the shots were fired.

Skipper saw the transition from one life and culture to another, and he and the people that surrounded him provided a link to a past frontier that is gone forever. Skipper's nephew, Ginger Gurrell, was perhaps the first hippie. A Harvard-educated recluse, he built a cabin in the mountains filled with books and records of opera and classical music. Artists such as Gollings stayed at the ranch and left sketches behind for the walls of the guest cabins. Priceless Native American artifacts decorated the Horton House, as Skipper's home was known.

Because they traveled a long way and transportation was slow, early guests stayed the entire summer, and from the start, the ranch reflected a relaxed and carefree attitude. Although problems, contretemps, and inconveniences existed, no one took them too seriously. The cabins acquired names other than their official ones. Roaring Forks became Roaring Toilets after the precarious state of the plumbing in the early years. Lost Cabin became Lost Weekend after it was inhabited by a bunch of bachelors who would ride out of sight of the corral, dismount, and break out the bottles. Honeymoon became Brookside after an actual honeymooning couple found the name too cute. Hackamore was called Hesperus after the wreck described in Longfellow's poem because it was perched precariously over the creek and would groan and sway in high winds.

For the peace of mind of the guests, the office boasted a sturdy combination safe. For the convenience of office employees, the combination was written on the ceiling above the safe. The billing system known as the "bad news" was accurate but slow. One guest from Peoria had ordered a bottle of scotch labeled as an "accommodation" on the bill. The charge with his signed order appeared on his bill the following year.

The Forest Service, then as now, had a limited budget, and trails had to be cleared. Skipper was a regular Tom Sawyer about getting the guests to do the work and making them feel they were having fun doing it. The South Fork Canyon trail was frequently obliterated or obstructed by spring floods, and guests often found themselves on an organized picnic in the canyon clearing the way. These picnics occasionally included dynamiting rocks. Fortunately, OSHA with its strict safety requirements was not yet around to spoil the fun.

Skipper and his first wife, Gertrude Butler, who died at a young age on the ranch, raised three sons, Jack, Bill, and Bob. All three were part of the ranch as children but drifted onto other paths as adults. All three died in the 1970s.

Skipper and his many friends lived in heady times and observed an ecological and cultural transformation from a frontier to a civilization dominated by the white man. Anyone who knew Skipper or his companions could not have helped but be fascinated by their stories of the shifting of worlds. As intriguing as this window into the past is, the story of this book does not start until Skipper traveled to Chicago for a press party as a national committeeman for the Republican party.

THE POSTMISTRESS

he man with the eight-foot bamboo fly rod assumed a comfortable, relaxed stance, looking down on the water with his left foot slightly ahead of his right. His right hand held the rod horizontally to the surface of the water, while his left held the line between the reel and the first guide. By lifting his right forearm and pulling the line down with his left hand, he began to pick up the line in a motion that gradually increased in tempo, so the fly at the end of the leader barely disturbed the surface of the water.

With his right wrist bent slightly, he prepared for the most critical part of the cast, the backcast. His forearm and the rod moved upward at moderate speed. At the end of this motion, his cocked wrist snapped up, throwing the line high and extending it well to the rear. The tip of the rod bent backwards in anticipation of driving the line during the forward cast. At the exact point of maximum height and full line extension, the man would begin his forward cast. If he started too soon, he would pop the line like a whip, too late and the barb of the fly could catch him in the neck or ear.

The forearm began the forward cast powered by a snap of the wrist. The cast was aimed above the target on the water giving the line an opportunity to extend. The fly dropped right on the target.

With an insouciance that belied the skill and precision required, the man repeated this cast in mesmerizing, fluid movements, dropping the fly each time on his intended target. Henrietta "Hank" Stuart watched this performance in awe along with several hundred equally entranced observers. A beautiful pool on Rock Creek teeming with rainbow, brown, and brook trout? No. Actually, this scene took place in an arena in Chicago hosting an annual sports show in the mid 1930s. The trout stream was a long artificial tank and the targets were small floating rings. Hank, a sometimes drama critic but, more often, whatever was required, had been sent by her newspaper to interview the flycaster.

After the respectful applause had died away and the crowd disbursed, Hank advanced with notebook and pencil to request an interview. The man happily complied and Hank unleashed her well-rehearsed opening question, "Where do you practice your art?"

The man looked at her with incomprehension and finally said, "Why right here, of course."

Equally taken aback, Hank said, "No. I meant where do you do your fishing?"

The man looked at Hank in wonderment and replied, "I've never been fishing in my life. Why would I permit some slimy fish to ruin my flies?"

At the time of the interview, Hank had no idea she would soon be living near the confluence of two superb trout streams in a state she had only seen through a train window. In fact, gazing from a window in the observation car at the desolate surroundings of Wyoming, she had taken comfort in knowing she lived in a civilized place like Chicago. Hank may have been incredulous that a fly casting champion had never truly gone fishing, but she was sympathetic. If going fishing meant visiting a place like Wyoming, she, too, was ready to forego the pleasure. Little did Hank know that fate has a sense of humor.

Hank grew up in Plainfield, New Jersey, the educated daughter of a marine architect from Maine, and always retained some mannerisms of the East Coast. However, her hallmark characteristics of humor and joie de vivre were apparent at an early age. As a child, she terrorized her neighborhood by setting off a huge box of fireworks. It seemed she could not wait until the Fourth of July.

I am not sure exactly how affluent her upbringing was, but she delighted in telling of an old family friend who would take her maid to the country club every day for lunch. The ladies, elderly, eccentric, and slightly potty, lived adjacent to the club, which made dinning there convenient. Finally, the house committee received enough complaints that the chairman screwed up his courage to ask her to refrain from bringing her maid to the club. If the request were not honored, she would face expulsion. The elderly lady was, in turn, shocked, hurt, and then enraged. She reminded the chairman that she had given the club a property easement that extended from her house across the seventeenth fairway and that it was renewable annually. After consulting with the County Registrar of Deeds, the chairman affirmed that the club would be delighted to have her grace dine at the club and with whomever she chose. In fact, the next meal would be complimentary.

It took a woman of some courage to pursue a career in journalism far from home in the 1930s, and Hank maintained an abundant supply of courage. She worked in radio for a time and had an amazing ability to time her writing to the second. Her soap opera dramas could be read in exactly fifteen minutes.

In the 1930s, radio was very much live time as there was no ability to prerecord. Consequently, lengthy rehearsals were required to reduce errors that could not be dubbed out. Sound effects had to be created on the spot. Hank wrote a story calling for a gentle rainfall. To convey the appropriate sound, Hank and her colleagues tried everything from crinkling paper and tin foil to dropping rice in a pan. Finally, a young page delivering mail asked, "Why don't you try water from that watering can on the shelf?"

Hank and her colleagues glanced at each other surreptitiously until one man said, "Beat it kid. What do you know? And close the door behind you." When the door closed, they simultaneously rushed for the watering can, tried it, and found the perfect sound of rainfall.

Hank met Skipper at a press party, which he attended as a national committeeman for the Republican Party. He invited Hank to Wyoming to view the ranch, but she initially demurred. The only thing she had seen of Wyoming was from a Union Pacific Pullman car, and she hadn't found what she saw particularly appealing. Love eventually triumphed, and she went to Wyoming with a carefully composed message in her pocket. She tried to send it to her editors in Chicago on the day she arrived, but the telegraph office in Buffalo was closed and the ranch had no phone. After several weeks, she returned to Chicago with the message still unsent. It read, "Send me a telegram telling me to return immediately." Hank and Skipper were married in 1937, and Hank joined Skipper at the HF Bar Guest Ranch, quite an adventure for a young woman from New Jersey.

Although she had a commanding and magnetic presence, Hank was not tall. Her face was oval and full featured and her hair was generally worn in a pageboy cut. Dressed in her customary blue jeans and weathered by sun and wind, she radiated an aura of class and elegance that did not need jewelry or other accouterments. Even in the corral surrounded by piles of manure, she retained an East Coast appearance that was perhaps augmented by the Brooks Brothers shirts she favored.

Although introverted and seemingly shy, Hank was bigger than life to those who knew or worked for her. She was amazingly phlegmatic and could take setbacks and disasters in her stride. She joined Skipper in the multilevel stucco and wood structure that everyone called the Horton House. Although no architectural marvel, it radiated charm because it soon reflected the personality of its newest occupant. From the Harrison's yellow rosebush on the verandah to the books, newspapers, western art work, and Native American artifacts, everyone could tell that an interesting person lived here.

During one extremely rainy summer, the ground became saturated, a rare event in Wyoming. An irrigation ditch that ran behind the Horton House was soon full to overflowing. Finally ceding to a dreadful thunderstorm, the ditch gave way and sent a torrent of water through the house. Priceless rugs and books and stacks of *New York Times* were washed through the house leaving mud and gravel in the wake. A disaster of this magnitude would undo most people. But Hank merely said, "We just lay the stuff out to dry."

Although not a native Westerner, Hank shared the sense of self-reliance that distinguishes the breed. She felt that virtually all maintenance could be handled by ranch personnel. Calling in and paying for a professional was a mark of failure, and if the truth were told, Hank could be a little tight about maintenance.

The cottonwood is the predominate tree along water courses in Wyoming, and the trees can grow to huge proportions. Because the cabins are nestled along the north fork of Rock Creek under cottonwoods, there is always the threat of a dead branch crashing down on an unsuspecting roof. And so it was not surprising that an extremely large and heavy limb eventually grew over a cabin with the unlikely name the House that Jack Built. The limb began creaking ominously, even in moderate winds, and Hank spent days surveying the obviously dead limb and discussing it thoroughly with anyone who could handle a chainsaw. Finally, biting the bullet, Hank called an arborist from Buffalo to trim the tree without damaging the cabin.

Why is it that professionals always look like teenagers? The arborist who greeted Hank looked very young and inexperienced. Although Hank did not ask to see his credentials, she did ask if he had done this before and if the branch could be removed without consequence. His reply was yesterday's equivalent of "no problem."

The arborist scampered up the tree and after tying off the limb, jerked the chainsaw to a roaring start. Fortunately, the house had been evacuated and there was quite an assemblage of onlookers eager to watch a professional in action. The saw bit into the wood, the ropes snapped, and the limb crashed into the roof of the cabin causing considerable damage. While this turn of events would have sent most people into a towering and justifiable rage, Hank said resignedly, "We could have done that."

Hank was plagued by a bad hip that she took in stride even though she walked painfully with the aid of a cane for a number of her latter years. Inevitably, in a ranch with rough ground and countless elevations, Hank took a fall and broke her hip, and an ambulance was called. A well-intentioned neighbor with a huge glass of red wine decided to comfort her by riding with her in the back of the ambulance to the hospital. There are two speed bumps on the property that are large enough to cause a well-loaded Buick to scrape bottom, and the gravel road frequently is no smoother than a washboard.

The two speed bumps sloshed an ample amount of jug red onto Hank accompanied by several "sorry about thats." The washboarded gravel transferred about half of the remaining wine onto her. Finally, Hank asked her solicitous neighbor to chug the wine or pour it out. She preferred to have the emergency crew think she tumbled from natural causes rather than be considered a falling down drunk.

When she awoke from one round of surgery on her hip, Hank was asked by the surgeon rather perfunctorily how she was doing. Hank glanced drowsily at her toes and said, "Other than the fact that one leg is now shorter than the other, I feel pretty good."

In a bit of political legerdemain that was probably Skipper's doing, the ranch was designated an official post office with all the perquisites appertaining thereto, including wanted posters and dire warnings about molesting post office personnel. These were all displayed with appropriate good humor. Naturally, a new post office needs a name. And what better way to come up with a name than to get the wranglers and other ranch staff together, let the beer flow, and have everyone begin to suggest names of animals and rhyming body parts. Two names were submitted. Fortunately, Deer's Ear was rejected and Saddlestring was the name accorded by Washington in its wisdom. Hank became the postmistress. Although Hank took her duties as an official U.S. government agent seriously, she also enjoyed the perks of the position. For example, when she received unwanted mail, she would simply stamp it "deceased" and return it.

Undoubtedly, because of her natural leadership qualities, Hank was also appointed the air raid warden for Sheridan, Wyoming, during World War II. I think Hank accepted because she was given a droll hat to wear and an armband. The duties of an air raid warden were not particularly onerous because even the most ardent booster of Sheridan could not have put the little town high on the enemy's priority list.

In case of enemy attack, the plan approved by the appropriate civil defense authorities was for the citizenry to jump into their cars, race up the Red Grade, and disburse into the forest. Unfortunately, cars and gasoline were in short supply owing to rationing. More important, the Red Grade, aptly named for the color of the dirt and gravel, was, and is, a narrow, rough road that climbs in tortuous hairpin curves up the foothills into the Big Horn Mountains. Frequent turnouts provided spring water for overheated radiators.

Knowing that if the bombs ever fell, the citizenry would be far safer in their beds than converging en masse on the Red Grade, Hank performed her duties as an air raid warden but kept the master plan to herself and a few friends with a good sense of humor. She did enjoy the hat.

In fact, Hank loved hats and costumes and kept a chest full of them in Horton House. Whenever a gathering reached a certain momentum, which usually meant that the libations had attained critical mass and a retired admiral was powering the piano with tunes that could only have been learned in a bordello, Hank would open the chest and distribute the hats and costumes. Pandora would have been proud. Only a person walking into the room stone sober found the hats unusual for people who were prominent citizens and highly respected in their various communities. Hank usually reserved the grass hula skirt for the moment when most inhibitions had run for cover, and the chosen wearer would perform enthusiastic bumps and grinds to a mixture of cheers and cat calls.

Hank's hat of preference was a multicolored beanie with a propeller on the top. When importuned by the assemblage, the piano player would strike up "Down by the Old Mill Stream," and the gathering would burst lustily and occasionally, tunefully, into song. With the propeller fluttering, Hank would pantomime the words, rotating her forearms for the mill, using a flowing motion of her hand for the stream, and performing other highly original, if not amazing, interpretations.

No matter how many times people saw this performance, it always brought down the house. The guitars would come out next, and the songs would range from tearjerkers like "Little Joe the Wrangler" to limericks that would make a mule skinner blush. Finally, people would say their long overdue good-byes and exit through the patio into pitch darkness. There were a number of stone steps descending to the road, and at that hour, the descent was always precarious. People would argue about how many steps there were and on which foot you should start and then launch themselves into space. The inevitable thuds, crashes, "Oh my Gods," and shushes followed.

By 1948 when Skipper died, Hank had learned to manage the ranch through on-the-job training. Hank said that Skipper let her make her own mistakes as long as she fixed them, and this approach proved to be a good management training program. Hank jokingly said that she was not born to be an innkeeper, and she did give some thought to selling the place after Skipper died but fortunately decided to carry on.

Hank always had an interesting perspective on even the most commonplace events. She preferred flying on smaller planes feeling they were safer than big airliners. She reasoned that people die when their number is up, and on smaller planes, there are fewer people whose number could come up. This kind of reasoning makes logicians and statisticians laugh condescendingly at first and then stare off into space muttering.

Hank considered declining to contribute to the restoration of the Sheridan Inn because she had trouble supporting a historic monument that was only ten years older than she was. She bristled at the sign at the ranch freeway exit that said "No Services Ahead." At the swimming pool, she once argued that the ability to float was more important and more difficult than swimming itself. When asked to demonstrate, she floated so proficiently that she fell asleep. When she woke up, everyone had disappeared.

Hank's driving habits were legendary. Anyone living in Wyoming will put many miles on a car, and Hank was no exception. Her vehicle of choice was always the largest stationwagon made, which in the 1940s was wooden sided. The car sported, as did every successive ranch car, a 16-1 license plate indicating it was the first issued in Johnson County. Such a plate is very easy to spot.

Hands clenched tightly on the steering wheel at the 10 and 2 position and eyes barely peering over the steering wheel, and, in later years, through the steering wheel, Hank would careen down gravel roads at death defying speeds. There was always a lot of movement between the gas and the brake pedal, so unsuspecting riders experienced a roller coaster–like, and sometimes nauseating, ride.

Younger, inexperienced, and as yet, politically naïve police officers would occasionally attempt to stop and ticket her. One such policeman followed her the entire thirteen miles from Buffalo with siren screaming and lights flashing. With the windows rolled up, the radio blaring, and eyes squinting over the wheel, Hank had no idea she was being followed. When Hank parked the car at the ranch, the policeman pulled up ashen faced and trembling and said, "Do you realize you were going eighty-five miles an hour on a gravel road?"

Hank replied, "That's amazing. I didn't think such a speed was possible." She then turned her back on the officer and walked into the office. About the only other time a squad car would appear on the ranch was when Hank's setter would go wandering and get into mischief. Then a policeman would drive up with setter happily pacing the rear seat and barking out the windows.

Hank often drove to Sheridan to pick up guests and staff, and the ride back to the ranch could be harrowing, silencing the most loquacious of passengers. Once, Hank drove a young guest to the top of the Red Grade, which can be a nerve wracking ride with the most sedate of drivers. Hank put on her usual performance—rapidly accelerating, decelerating, braking, shooting around blind curves with an occasional warning honk and drifting perilously close to unguarded edges.

At the top of the Red Grade, the young passenger asked if Hank would stop briefly. Hank complied, thinking how nice it was for a young man to be so aesthetically oriented that he wanted to enjoy the spectacular view. The young man stepped out of the car, vomited, returned to the car, and said, "Thank you." Somewhat nonplused, Hank could only reply, "You're welcome."

Hank was a very kind person and never said anything to put people down. When pushed to extremes, she might refer to someone who was rude and cheap as "not very expensive." The Wyoming Chapter of the Colonial Dames and the Daughters of the American Revolution might be called "the pink hats" or "the mink and manure set."

Although an excellent horsewoman, in the early 1950s she decided to give up riding. She had ridden to one of the hills linking Castle Rock and the Saddle and had dismounted to pick wildflowers. Glancing up, she noticed a bear slowly working its way toward her and her horse. The scent of a bear will normally put a horse in the next county leaving the rider to deal with the problem at hand. Fortunately, the horse had not picked up the scent and was grazing peacefully with reins trailing. Hank remounted and returned to the ranch. Although in no real immediate danger, the incident caused her to think about being stranded far from home with a bad hip. She never rode much after that.

During the winter, Hank would say that she got terribly homesick meaning she was sick of home. At the first sign of snowfall, she would reach for an atlas and go anywhere warm and preferably exotic. On one of her trips deep into Africa, she was on a river boat tour that included a group of pink hats. They debarked while some scantily clad native porters unloaded their bags in the intense heat. As one would expect, there was some confusion in getting the correct luggage to its appropriate owner that generated considerable and incomprehensible discussion. One of the pink hats berated a struggling porter by saying, "No, no! Not the Louis Vuitton. The Gucci." Hank was incredulous.

Hank was fond of vodka and learned on one of her early trips to South America that it was not easy to obtain good vodka in that continent. Having tried a local brand, she left it in her hotel room when she checked out, thinking that someone on the hotel staff might find it potable. As she was walking out the front door, a room maid raced across the crowded lobby shouting her name and waving the bottle of rot gut. Red faced, Hank retrieved the bottle while the maid stood her ground expectantly with extended palm. Hank tipped her leaving a lobby full of bemused people. Hank tried to leave the bottle in her hotel room in the next city, and the exact same scene was repeated. Totally embarrassed and extorted out of two tips, which exceeded the cost of the vodka, Hank left the bottle in the taxi and fled the country.

Hank's loyalty to her friends was never questioned. On his deathbed, a long-time resident told Hank that he would like his ashes spread over the high mountains, not an easy task if you want to achieve real dispersion. Clearly, a small plane was needed. Hank attended the funeral service in a somber navy blue suit and then drove to the airport with the ashes. The pilot took her high over the Big Horns to the appropriate altitude and location. Hank slid open a window, dumped out the ashes, and they all blew back into the plane onto her navy blue suit. We can only hope that a few ashes hit the desired target because most ended up in a vacuum cleaner or the Buffalo laundry.

Tom Baird spent many an evening in Hank's living room chatting amicably about issues as wide ranging as their diverse interests and backgrounds. One year, Tom was at the ranch during the offseason doing some research for a novel. While chatting with Hank, the phone rang and Hank asked Tom to answer. Tom dutifully complied saying "HF Bar Ranch" into the receiver. A voice apologized for a wrong number and hung up. Sensing a unique opportunity, Tom carried on a one-sided conversation as follows:

"Rome?" Pause.

"Il Vaticano?" Longer pause.

"Why yes it is, Your Holiness." Very long pause with intently furrowed brow.

"Beatification? No, I'm afraid you have the wrong Horton."

My uncle Truman always thought Hank was one of the most fascinating and fun-filled women he had ever known. On one typical adventure, they went into Buffalo to watch a major parade celebrating an important anniversary of the town's founding. Truman put on his elegant butterfly cowboy shirt and multihued boots that Hank referred to as his "pimp pumps." Hank even wore a dress. It was that kind of occasion.

The parade committee discovered that the great great whatever of one of the invaders in the Johnson County Cattle War was now a prominent banker in Omaha and had the same name as his forebear. He was invited to ride a horse in the parade and, perhaps, be hanged in effigy, all in good fun, of course. As feelings concerning the war on the Powder River were still on smolder, the banker politely declined, citing urgent business in the opposite direction.

Hank and Truman sat on a curb by a bandstand and watched a marvelous flow of costumed participants on horses, wagons, floats, and foot. The Basques were prominent in the parade keeping alive a hundred years of tradition in Wyoming. Originally from their own region of Spain, they had become first expert sheepherders in the state and later, prominent citizens. When I was a very young boy, I sat in a Basque sheepherder's wagon on the top of the Saddle and heard the end of the war in the Pacific, and thus World War II, announced on his radio.

Hank and Truman watched with mounting interest as a bunch of sheep were herded down the parade street by a watchful Basque and a hyperactive border collie. A good shepherd and dog working sheep is beautiful to behold and this exhibition was no exception. However, working sheep in the middle of town in a parade environment is not like the open range. The normally tranquil shepherd had a look of desperation, and the border collie appeared to be on amphetamines.

If someone could have viewed the event from a more global perspective, the disaster might have been averted. However, spectators sitting on a curb right next to the bandstand with their eyes glued on the sheep, did not sense the impending disaster. The band director had given the musicians a well-deserved break and was now ready to begin the pièce de résistance. His back was turned to the parade and his mind was on music, so he was totally unaware of the approaching sheep.

Just as the sheep were abreast of the bandstand, or ground zero in nuclear terms, the band director gave the down beat and the musicians exploded into "Star and Stripes Forever" with an enthusiasm that bordered on ferocity. John Philip Sousa would have been proud of the power of the trumpets and tubas. Of course, the sheep exploded as well, plunging into the crowd searching for any avenue of escape back to the open range. The sheepherder was aghast and the border collie went manic.

After dodging flying sheep, Hank and Truman rolled helplessly on the pavement in laughter. They were so overcome by this event that they decided to return to the ranch to recount the story over cocktails. This decision caused them to miss the balloon ascension.

A balloonist, claiming to be a descendant of the famous Monsieur Picard, was to carry the mail over the Big Horns to a post office on the west side of the mountains. This was an ill-starred venture to begin with because the prevailing winds blow to the east, not the west. Nevertheless, that day, conditions seemed propitious and the balloon lifted off and headed west. Hank and Truman unknowingly streaked ahead of the balloon and were well into cocktails recounting the sheep incident to a growing audience. Suddenly, an awed gasp startled the group and everyone looked to the south to see a huge balloon descending into what was the dump at that time.

For someone with a bad hip, Hank took off like a greyhound and ran a direct course through barbed-wire fences. By the time she reached the downed balloon, her dress was in tatters. The remaining party goers came panting forth to see Hank demand the mail from the bewildered balloonist who asked plaintively, "Who are you?"

Hank replied, "I am the postmistress of Saddlestring, Wyoming."

Running a nine thousand acre dude ranch with thirty-six buildings and associated plumbing and electrical issues plus a regular staff and about forty-five seasonal employees takes more than one person. Fortunately, Hank grew to rely on people like Dean Thomas.

DEAN, HORSES, AND PLUMBING

n late June 1951, a fifteen-year-old kid got off a Western Airlines plane in Sheridan and walked hesitatingly to the terminal. His mother had assured him that he would be met, but it was about 10:00 p.m., pitch dark, and the plane was several hours late, a fairly common occurrence for the Sheridan airport. A touch of anxiety swept over him as he thought about no one being there and how he would arrange transportation to the HF Bar Guest Ranch at that hour when there did not appear to be a taxi stand. Completing the worst-case scenario, he wondered if someone were there to meet him, how they would recognize each other.

He straggled into the airport with an armload of miscellany including a tennis racket as his mother had mentioned there was a tennis court at the ranch. Although a court once existed, I have never seen it in use. The freezing and thawing of Wyoming winters and the relentless sun of the summers can make quick work of a tennis court. I am not sure what optimist installed the court in the first place. As competent as cowboys are at fixing things on a ranch, tennis court maintenance is not their forte. At one point, the court was screened over, and there was an ill-fated attempt to raise game pheasants. I think the remains of the court have returned to the soil by now.

Spotting anyone at the Sheridan airport is not difficult, but when it is a scared fifteen year old carrying a tennis racket, even the visually impaired could make the connection. A cowboy stepped out of nowhere and said, "Hey, you must be Freddie's boy. I'm Dean Thomas." Those were the only words spoken for the next hour.

Forty-six years later, the "kid" described the encounter as follows:

There was a handshake, and I recall a square, rough, very strong hand. That was followed by a wait, in silence, for my luggage and a faintly quizzical look at my tennis racket and, when they came, two large suitcases. I translated the look as, "Real men need no more than a bedroll and a small valise, and they certainly don't play tennis." Talk about intimidation. Here was this trim, forty-year-old cowboy with movie star looks wearing real cowboy boots, blue jeans, a leather belt with a big silver buckle with "DEAN" carved in the back, and a form fitting snap shirt. I'd never seen a snap shirt before. The hat was set just so, as was the Camel unfiltered cigarette there in the corner of mouth. He wasn't actually very tall or heavily built; he was compact, extremely strong and, to my eye, without any apparent effort, absolutely radiating power. He might as well have been seven feet tall and weighed three hundred pounds.

When I last saw Dean, perhaps in the late 1980s when I was fifty something, he still had the power, merely by cocking his head and giving me a sideways glance from under his hat brim, to send me into a total funk. Surely I had said or done something wrong or dumb or had forgotten something. Now that's intimidation.

We then walked out to the car, where, as I was putting my bags in the back of the huge, gray Chevrolet stationwagon, I noticed the 16-1 license plate. I was impressed and turned to ask a question of Dean, but he was twenty feet away, taking a leak against the side of the terminal. I realized that cowboys could even pee with panache.

I have few memories that are as sharply etched as these first few minutes with Dean. There are few people I respected more, and none I feared more or whose respect I sought so assiduously. I like to think that, somewhere along the line, I was, at least, modestly successful in this last regard.

The need to pee against the side of the terminal was natural, because when Dean had learned that the plane would be late, he reflexively adjourned to the Mint Bar, which is the logical place in Sheridan to kill a few hours and enjoy some Pabst Blue Ribbon beer or "blues" as they were called by the regulars. Blues were the cowboys' beer of choice, although most ranch guests drank Sheridan Export, which was exotic because it could only be obtained locally and was not exported anywhere. For generations of young men and women, Sheridan Export was our very first beer, and it was usually purloined from our parents' iceboxes. Although the brew does not survive today, many think of it nostalgically whenever they drink a beer from a microbrewery.

Dean was born October 23, 1910, and grew up in the Sheridan area. He started working at the HF Bar in 1931 and like every young man, his life was interrupted and changed irretrievably by World War II. In its infinite wisdom, the army assumed that everyone who was drafted in Sheridan and Johnson County would know how to pack animals.

Consequently, Dean and his friends were sent to an army packing school where they had to unlearn everything they already knew and begin packing mules the army way. In later years, Dean exhibited a great deal of respect, and even fondness, for mules, which was strange as familiarity in the army often produces the opposite effect.

Apparently, the army did not think that the Germans could be defeated with only mule packing aptitudes, so the Sheridan inductees were sent to Colorado to learn winter survival skills. In fur-lined chaps, caps, and overcoats, Wyoming cowboys have survived winters quite adequately for over a century. In army-issued winter gear, the inductees almost froze to death. Dean said he had never been so cold before or since in his life.

By all accounts, the worst part of the training was learning how to ski. Although Dean was well coordinated and athletic, try as I will, I cannot imagine this bow-legged cowboy swishing down a ski slope. Apparently, most recruits were somewhat skeptical of their ability to learn this art, and perhaps it was this defeatist attitude that caused such a disaster.

Army ski equipment was little better than boards strapped to boots, and the accompanying instruction was neither subtle nor elucidating. Dean described a line of freezing recruits at the top of a hill being told to descend when the signal was given. Most launched themselves unquestioningly at the appropriate time, and the recalcitrant, and perhaps rational, ones were pushed.

The smart ones fell down before they reached bone shattering speeds. Because the bindings were little more than straps, it was virtually impossible to maneuver, so many soldiers just crashed into each other. Anyone who got beyond the initial mayhem could only pray that he was heading in a direction without obstacles. As the injuries mounted, the army concluded that the cowboys had been converted to maximum readiness, and they were shipped overseas.

Dean was sent to Italy and did pack some mules into that mountainous terrain. Apparently, the mules served admirably and complained much less than the men. After the war, Dean returned to the ranch, and a number of his fellow soldiers from the Sheridan area were hired by Tee Pee, a guest ranch located up the Red Grade and into the Big Horns. They were first-class hell raisers, and Dean described one of them as the ranking second lieutenant in the army because he had so many infractions of regulations and other contretemps that he never made first lieutenant. Having been a first lieutenant in my youth, I know that making that grade is no trick as long as you show up.

Although desperately short of help, Tee Pee finally had enough of their antics and fired them en masse. The HF Bar was short of help as well, so Dean hired them all. He claimed he was able to get a reasonable day's work out of them most of the time but did admit that it was the wildest summer he had ever spent.

Sometime after Skipper died in 1948, Dean felt there was not enough work to keep him occupied, so he joined the T Bar H Guest Ranch in Arizona. The clientele included a number of Hollywood types, and Dean got along famously with the likes of Clark Gable and Groucho Marx. According to Dean, Groucho was a proficient equestrian, or as he put it, "He rode pretty good." In 1956 Hank wrote Dean from Europe asking him to return to the HF Bar, which he did, and he remained for the rest of his life.

Although Dean had many talents, the HF Bar dudes thought of him primarily as the corral boss, and certainly, horses did occupy a significant part of his time. Because of constant attrition through age, injury, or winter, replenishing the herd was a constant challenge. It was in this capacity that Dean met the original Marlboro man who was a horse trader from Riverton. Ironically, they both died of lung cancer. Over the years, Dean knew about everyone in the horse business and was always able to find animals, even when others could not.

After the war, a number of military horses came on the market, and Dean snapped them up along with surplus saddles and other tack. The horses were generally good stock, reasonably well trained, and could be distinguished by numbers branded along their necks. The saddles had high horns and cantles with cartridge pockets attached. They have long since been replaced with more streamlined versions. Kings in Sheridan has been selling excellent saddles and tack for years and maintains a marvelous museum illustrating changing fashions over time.

The price of horses can fluctuate significantly depending on quality, fads, and supply and demand. On one end of the spectrum, the canners and cutters set somewhat of a price floor as horsemeat continues to be a main ingredient in dog food. Dean had also been known to deal in that end of the spectrum because some of the guests preferred a slow moving animal with low elevation in case of a fall.

Dean occasionally bought retired race horses, but the results were never optimal. Because of their rather narrow purpose training, they were not particularly sure footed on mountain trails and did not seem to get along well with anyone, man or animal. They were also high spirited and incredibly nervous. Nevertheless, the racetrack was one of many unusual sources of animals that Dean scouted and frequented.

For many years, the ranch ran its own cattle or leased pasture to others. To conserve forage, they moved the entire horse herd to a box canyon leased from a rancher in Montana. This drive was one of two great work and social occasions for the ranch staff. The horses were driven north along the highways, or down to Montana as the old timers say, sometime in middle to late September. The wranglers worked pretty hard but made a good party of it, before, during, and afterwards. They then repeated it in reverse the next June. The drive was discontinued in the mid 1950s as there was no way to run horses across Interstate 90. Such is progress.

The other great social occasion was the gathering of ice from a pond on the adjacent UM Ranch that was once owned by Skipper as was the Paradise Ranch on French Creek high in the mountains. Unfortunately, both properties were jettisoned during the Great Depression. The ice was used to fill the ice boxes located on the porches of the guest cabins. There is still an ice box on every porch, but today the boxes are filled with artificially and hygienically manufactured blocks of ice for cooling drinks and making ice cubes. Before the ice making machine, Dean and others would pack a picnic lunch, plenty of libations, and a big tree saw and head over to a frozen pond on the UM. Cutting blocks of ice was heavy work, but the picnic and libations made it a great social occasion. The ice blocks were packed in an ice house and covered with the same sawdust that was used to make the "dynamite" or kerosene-soaked sawdust used to start log fires in the cabin fireplaces. Consequently, the ice occasionally made a good scotch taste vaguely like an accelerant, but no one seemed to mind, and the ice generally lasted through August. A well-intentioned and newly arrived health inspector put an end to the ice gathering party.

No one will argue against interstate highways and health inspectors, but they did wreck havoc on long standing traditions. One tradition, about which we never learned enough, concerned the railroad trips Dean would take on the Burlington when he accompanied HF Bar cattle to the stockyards in Omaha.

The South Omaha stockyards were once highly competitive, and despite the distance, an owner could occasionally profit by shipping cattle to that market. The steers would be loaded onto cattle cars, and someone had to keep them standing so they would not be trampled. Apparently, the words *cowpoke* and *cowpuncher* were derived from that chore. Loading cattle and keeping them standing was hard, grimy work, but at the end of the rail was the Castle Hotel where cowboys, ranchers, buyers, and providers of other services gathered. Dean never went into much detail about his activities in Omaha, but whenever he mentioned the trips, his eyes took on a faraway look, and a little smile appeared as he withdrew into silent reverie.

One of Dean's many challenges in the corral was assigning horses to guests of wildly different, often overestimated or unknown, riding skills. The assignments often gave the horses attributes they could not possess in a million years. One experienced guest brought her grown daughter to the corral and assured Dean the young woman was an excellent rider. The daughter was quite tall, so Dean brought out a huge and spirited animal. A wrangler saddled the horse, and the woman hiked a left foot to the stirrup, bounced several times on her right foot, and launched her right leg and remaining body with great gusto over the high cantle. It would have been a mounting demonstration worthy of a western film stuntman except that her momentum was so great, she somersaulted off the other side of the horse and landed on her back in corral dirt and excrement.

Dean watched this performance with a barely perceptible widening of the eyes and then suggested that a shorter horse was perhaps more appropriate. He roped a small, lethargic animal who followed him docilely to be saddled. Dean assured the woman that this was a horse of great heart and stamina and would be perfect for someone of her experience. And he was.

Later in her stay, the same lady mounted up wearing a poncho as the day looked threatening. She had unwittingly forgotten to zip the fly on her rather baggy blue jeans, and as she launched herself into the saddle with her usual élan, she snagged her fly on the saddlehorn. This movement was followed by wild and erratic gyrations, the cause of which was hidden underneath the poncho. She finally extricated herself and moved out of the corral as Dean looked on incredulously. He shook his head and muttered something about horsemanship in the Washington, D.C., area.

In the summer of 1961, a young couple was working at the ranch, one mowing hay and the other managing the dinning room. The parents of the young man were at the ranch and were taking their usual pack trip into the mountains. The couple arranged a few days off to join them at Seven Brothers, the glacially formed lakes in the alpine region. The lakes are absolutely magnificent and well worth the long and arduous ride from the ranch. Dean felt a little more horsepower was needed for the trip and brought out a large paint named Skunk for the young woman.

Most horses are fairly gregarious and like the company of other horses, so keeping a riding party together on a trial is usually not difficult. However, if horses on a trail ride will not stay together naturally, good horsemanship demands that you stay caught up with the rest. If one horse has a slower walk than the others, the horse's rider is periodically forced to push the animal into a trot. Cowboys can trot for hours without their butts leaving the saddle. How they do it is a total mystery, defying all logic and laws of physics. Dudes are invariably jarred and bounced, and after several minutes, the pain can be excruciating.

Like people, every horse has a different gait, and like Chevrolets and Mercedes, some are smoother than others. Skunk had to have the roughest ride of any horse in Wyoming, if not the Northern Hemisphere. In addition, Skunk saw no reason to remain with the rest of the horses, so his rider was forced to repeatedly urge Skunk into a trot. After a ten-hour ride, the poor woman could barely walk or stand and was convinced that this was Dean's way of expressing disapproval of her leaving her dinning room duties.

After two days with the group at Seven Brothers, the pain had eased somewhat and the magnificence of the surroundings acted as a balm. The sparkling Seven Brothers lakes are nestled under glacial cirques with sheer granite walls shooting straight up and then disappearing into mounds of dazzling snow. In the presence of such overpowering beauty, one tends to forget muscular discomfort.

Unfortunately, on the return trip, the pain resurfaced with a vengeance and untold new torments were added. In fairly typical fashion, Skunk was a much rougher ride going downhill and over four thousand feet in elevation had to be reversed. The last insult was the descent of Stone Mountain during which each painful lurch of the horse caused the young woman to conclude decisively that Dean had planned the whole thing.

A little after Red Canyon, she dismounted and walked the rest of the way to the lane leading to the corral. There she remounted and rode into the corral where the entire crew was waiting.

Without preamble, Dean asked, "How was Skunk?"

The young woman's nonchalant reply was worthy of an Academy Award, "Great. Couldn't have been better." She gritted her teeth, strode to the cabin, and collapsed.

Was this incident planned? Thirty-six years later, the husband writes,

> I initially expressed doubt that Dean would do such a thing, but as I thought back over the years, I wasn't so sure. I recalled that when they deemed it appropriate for any of a number of obscure reasons, Dean, Barney, Bob Ross, and others, with the barest hint of a smile and talking in the cowboy code that only other cowboys can decipher, would load some poor soul on an absolutely inappropriate horse. The big-talking, know-everything, "expert" macho horseman might just as well find himself on Dynamite, a real bronc that would only run, as on Jasper, a plug who could only walk. Nothing was ever said about this, and although one can never be positive about such things, I think it likely that my wife was the butt of some cowboy humor. By not complaining, she thinks she fooled Dean. My guess is that Dean knew better.

When guests would go on a pack trip, generally to the Willow Park cabin at about eighty-five hundred feet into the mountains, a standard ritual would unfold. Incidentally, the real Willow Park was, in fact, a beautiful park, or open space in a forest, located in a small valley with a meandering stream enshrouded with willows. HF Bar had a large cabin on the creek with a lofty pitched roof and a porch filled with firewood. The fishing was spectacular. As a boy in the late 1940s, I remember riding up the valley and stopping to visit with members of the Army Corps of Engineers. The engineers would talk vaguely about a dam but seemed to be doing as much fishing as surveying. We all thought nothing would happen, but of course, it did. The dam was completed, and today Willow Park is a lake so that water can be diverted from the Piney Creek to the Rock Creek watershed. The Corps of Engineers built an open-faced sleeping lean-to and a cooking cabin in a beautiful spot further up the creek for the ranch. Although these cabins are now nestled in the pines, they are still called Willow Park.

The mountain packing ritual would begin with Hank who would discuss and arrange the menu. She would describe a wide variety of mouth watering options, but the final result always seemed to be the same and quite good nevertheless. The selections generally started with steak and other items that would be consumed immediately and then ran to canned goods, as there was obviously no refrigeration. Hank always enthusiastically recommended the canned chicken fricassee, but of course, she was down below and did not have to eat it. My nephew who was in single digits at the time, unwittingly, and perhaps prophetically, called it *frick'en chickassee*.

On the day of the departure for the mountains, the guests would deposit their gear in front of the barn where several pack horses were tethered. Then Dean would work his packing magic. After much grunting, squinting, mumbling, and repacking, Dean would pronounce the job accomplished and off they would go. It usually took about an hour to climb to the Saddle and by then, the packs would be slipping terribly with gear occasionally dropping and causing frequent stops for retrieval.

At the top of the Saddle, and out of sight of the corral, the mountain man would resignedly repack the horses. Instead of a diamond hitch, Dean must have been using a zircon configuration because the same ritual was repeated year after year. We never told

Dean and can only guess that the army mule training ruined his packing abilities.

Dean was extraordinarily handsome with an engaging personality and ear-to-ear smile. Most women found him extremely attractive and some pursued him relentlessly. The corral is located on a slight elevation so anyone approaching it can be viewed quite easily, especially from horseback. If Dean wanted to avoid someone in hot pursuit, the wranglers had a warning code. "Let's saddle up ole Strawberry" might send Dean off to the barn or up into Wranglers' Roost, the bunkhouse. On the other hand, "Let's bring out ole Wild Flower" might start him preening.

I doubt if there are many cowboys who ride consistently that have not broken a bone at some time or other. Horses spook at even the most common sights. To that little brain, a log or a rock can suddenly become a pouncing mountain lion. Even the most skillful and careful rider cannot always avoid a fall from rough terrain or slippery mud. Dean was no exception to the rule and once broke both wrists simultaneously. Even immobilizing casts on both wrists did not dampen his spirits, and he described the aftermath by saying, "You really find out who your friends are."

Any rancher, dude or otherwise, must understand a lot more than horses and livestock. The general maintenance is daunting and prohibitive unless accomplished through self reliance. Although Dean understood farming, irrigation, light vehicle repair, road grading, carpentry, and welding, his admitted fortes were electricity and plumbing. The electrical wiring all over the ranch was bizarre and would not have passed code in Somalia had there been any inspectors. It was generally quite visible and problems such as molten copper could be easily spotted. The plumbing, however, was an often invisible Rube Goldberg tour de force that over the years evolved into its own unique character.

After years of patching, burying, spot welding, and general tinkering, Dean was the only person who knew all the plumbing system's intimate secrets such as the location of certain pipes and shutoff valves. Dean did not share his knowledge of the system and often referred to the plumbing as his job security.

The guests often inadvertently inflicted damage that would inconvenience everyone. For example, in the electricity department, someone once brought an electric blanket that shorted out the entire ranch every evening. The system was so delicately balanced that no one, including the perpetrator who was asleep at the time, could figure out the cause. When the guest left, the problem ceased, and Dean declared the wiring sound. It was not until later that the connection was discovered.

As an example in the plumbing department, the main water pipe was located quite close to the road by the cabins, and a huge, protruding shutoff valve was unfortunately situated by a blind curve where the road turned at a tight angle around a cabin. For years, the valve survived unscathed until the summer that Admiral Greer arrived. Dopey, as his close friends called him because he was just the opposite, might have been highly proficient at docking a destroyer. However, in the car department, he was sometimes distracted. He soon roared around the virtually blind corner and hit the valve sending a surge of water high into the air.

After shutting down the entire water supply for the ranch, Dean spent most of the day trying to get a workable weld on the valve. The admiral was thereafter referred to as Geyser Greer. Somewhat later that same summer, a man who was an excellent driver and races cars as a hobby hit the same valve with the same results. Fortunately, the third accident that summer was caused by a ranch hand. The huge draft animal pulling the wagon that delivered ice and firewood (Hank referred to the combination as Ben Hur) was normally hobbled between stops. However, that summer some genius thought that a weight dropped as an anchor would be much more efficient than hobbles. Something inevitably spooked the horse, and he took off like a Triple Crown contender. The cart and weight barely impeded his flight. The flying weight bashed car after car parked along the road, but the real damage was done when the iron wagon wheel hit the valve with the usual and predictable results. By this time, Dean was beginning to lose his sense of humor. Although a major alteration for an amateur plumber, he moved the valve to a safer locale.

Dean had an uncommon, but pragmatic, approach to plumbing. A guest informed Dean of an overflowing toilet that could not be stopped and was flooding the bathroom. The bathroom floors in cabins that may have been started around the turn of the century are not always plumb. In this case, over an inch of water was accumulating on the floor. Dean sloshed into the bathroom, fixed the faulty plumbing, and then sloshed out to his truck.

At this point, the guest assumed he would return with a mop and a bucket. But Dean was not your ordinary plumber. He returned with a brace and bit, surveyed the floor with a critical eye, and drilled a hole in the correct spot. As the floor drained, Dean sloshed out, returned with an knife and a piece of wood. When the floor was dry, Dean whittled a plug, popped it in, and declare everything as good as new.

In addition to his accomplishments with horses and plumbing, Dean could naturally charm most people he met. Despite its geographic size, Wyoming does not have a big population, and Dean seemed to know a good portion of it. As an occasional team roping contestant, he knew many of the famous rodeo stars of his era and consumed many a drink with them. He spoke fondly of Tex with whom he would go on an occasional bender. Among other things, Tex was a brand inspector for the state. After a horrible car accident, Tex was permanently bent over at the waist. Consequently, he could not straighten up enough to read brands on taller animals. Today, this would be a clear case of disability. The state of Wyoming accommodated his infirmity by transferring him to the sheep inspection department.

Dean got along famously with the guests at the ranch, which tended to offset some of Hank's natural shyness. A life-long bachelor, he had, nevertheless, a special rapport with children, which is probably not that surprising since he did not live with them. Most parents of toddlers have a picture of Dean riding up and down the lane with a kid laughing and smiling sitting double in the saddle.

Dean had a natural ease with guests and did not worry about differences in education or incomes. He occasionally referred to the rich and famous as the "high poloi." Nevertheless, they found him fascinating and sought his company. He used to play poker with Potter Stewart, a guest who was at the time an associate justice of the Supreme Court. Dean referred to him as a "great guy and a damned good poker player" who regularly cleaned him out. After one frustrating evening when fate never dealt him a decent hand, Dean jokingly suggested that Potter's son, who was called Potsy and was six at the time, must have been flashing his dad signals as he was careening about the room. Actually, Dean may not have been joking.

Dean was a natural artist, and many of the cabins had charcoal drawings of horses and other western scenes over the fireplaces. These drawings were often produced during raucous parties that transformed guests into art critics egging him on with "helpful" suggestions. Dean once went to New York to take some art classes but concluded that they could not help him with what he wanted to do. Although the charcoal drawings have succumbed to general maintenance, the ranch still sends out an annual Christmas card featuring one of his drawings.

After a number of years of battling lung cancer, Dean died at the ranch on October 27, 1987. During his final day, he propped himself up in the back of a pickup truck so he could join others on a deer hunt.

JACK AND TRUDY

he little boys and girls pulling decorated wagons and carts or skipping and running across the lawn in front of the ranch house ranged in age from about five to ten. Although a first glance suggested they were off somewhere for organized play, a closer inspection revealed that they were imitating a rodeo parade. In fact, a rodeo queen led the group of floats, clowns, and contestants, some carrying homemade flags representing unknown territories.

After watching this fair imitation of the parades preceding rodeos in Buffalo and Sheridan, the bemused adult would rightly conclude that the instigators were Jack and Trudy Horton, more likely, Trudy. Naturally, a reader might guess that Trudy would be playing the role of the queen, haughtily leading her charges along the proscribed route. Actually, Trudy played many roles, none of them the queen. At times she was a contestant waving a banner. Or she might run up and down the line giving orders and holding the group together. Or she might drop by the side to be an enthusiastically applauding spectator.

If not Trudy, then who was the queen in the lead position? Actually, it was Jack made up and costumed by Trudy who had more important things to occupy her talents. What kind of power and leadership must a person have to dress up an older brother ranch kid as a rodeo queen and then organize an unruly bunch into a parade? Whatever it takes, Trudy had it in abundance as did Jack.

Skipper and his first wife, Gertrude Butler, had three boys, all of whom were raised on the ranch. Bob eventually moved to California and lived there until his death. Bill was quite a rodeo star, always winning several events at the Sheridan rodeo and others as well. Although twice married, he had no children. Jack, who was born on

October 18, 1912, attended Sheridan High School and then went to Pomona College in California. In 1931, he transferred to Princeton where he was a member of the polo team, a skill he picked up in Big Horn, near Sheridan, which has a long standing polo tradition. He probably did not get the idea of Princeton from Skipper who went to the University of Chicago. Jack married Josephine Jahn from Milwaukee in 1933 and graduated from Princeton the following year.

Jack became a geologist and worked for petroleum companies all over the world. Jack Jr. was born January 28, 1938, in Sheridan and Trudy in 1943, and the family lived in Venezuela for a number of years. As an unfortunate precursor of events to come, Jack's wife died of cancer before she reached thirty. Shortly after the end of World War II, Jack Jr. and Trudy went to live at the ranch with Hank, their step grandmother, while Jack Sr. led a somewhat nomadic life with the oil companies. Never comfortable with the western lifestyle, Jack and his second wife, Eleanor, would return to the ranch for several weeks as guests in the summer but then would be off traveling the world.

Jack and Trudy grew up on the ranch and attended a little one-room school house on Rock Creek, about a six-mile round-trip horseback ride. The remains of the school house are still standing. The brother and sister were remarkably similar. Although I have never really understood the term "rawboned," Jack defined it for me. He was tall, lean, and athletic with large hands and feet. Trudy was lithe with the unconscious fluid movements of a natural athlete. Both were highly intelligent and quick witted, and each had an ingrained curiosity. They were both somewhat introverted in the psychological sense of deriving energy from within as opposed to through other people. If other members of the ranch staff got together for beer and songs, Jack and Trudy probably would not be there. Far from appearing aloof, this introversion enhanced their magnetism, and their contemporaries were even more attracted to them. If there was a significant difference between the two, it was that Trudy accomplished her objectives with more visible enthusiasm and exuberance than Jack, who was equally effective but in a much more understated and laid-back way.

Trudy was highly competitive. When my cousin and I were about ten, we found an old shoeshine stand in the toolshed, so we started a shoe and boot shining business with the catchy motto, "You scuff 'em, we buff 'em." Sooner or later, all footwear on a ranch is covered with dust, mud, manure, or worse, so the money started rolling in, particularly on the evenings when dances were held. With a new source of discretionary income, we became pretty big spenders at the candy counter in the ranch store and soon attracted Trudy's interest.

Typical of Trudy, she started a business that not only made money but had a positive effect on the environment as well. In the days when the spaces were truly wide open, people seemed to feel that dumping a little refuse would have negligible impact. When you travel in the high country, you often come across caches of trash left by prospectors, hunters, loggers, and the military. Although eyesores, these dumps do have an archeological fascination if you are interested in arcane subjects such as what type of canned goods miners preferred a century ago. Fortunately, a "pack it out" ethic now prevails, although the Forest Service rangers, who have to pack it out for those who do not, would probably disagree.

Fifty years ago, some people thought nothing of tossing empty bottles from their porches into the creek, and many of these bottles were worth several cents in deposits. Dressed in tennis shoes and shorts, Trudy would skip from rock to rock collecting bottles in a huge sack. She could make a run from Frontier, the furthest cabin, to the store in no time at all, sort the bottles into trash and returnables, and cash in.

Trudy made the business look so easy, I thought I would horn in and diversify my sources of revenue. After all, she was several years younger than I was. I donned shorts and tennis shoes, slung a sack over my shoulder, and plunged into Rock Creek to seek my fortune. Less than 100 yards downstream, I emerged, scratched, bruised, and exhausted. The sedentary ways of the shoeshine business took on a new allure. To this day, I am not sure how Trudy did it. Maybe she could walk on water when out of sight. She was incredibly athletic and occasionally would do things such as walk to Willow Park carrying nothing more than a sandwich and return before dark, thinking nothing of it.

Trudy could always develop a rapport with people who were different, yet interesting. She was one of the few close friends of Ginger Gurrell, Skipper's Harvard-educated nephew who lived as a recluse. Ginger had a truck outfitted with a bed in which he tended to sleep whenever the hour or the mood overtook him. When he was in town, the mood frequently struck in the vicinity of a bar. He was somehow able to drive that truck over Stone Mountain and into the higher elevations, a feat which still seems impossible to me.

He built a cabin in a spot now known as Ginger's Meadow, and when the Forest Service removed it, he built another one further upstream on the south fork of Rock Creek. He assembled a water-driven generator to provide electricity for the cabin. He was an excellent photographer and took a number of marvelous wildlife photos using trip wires.

A man like Ginger is bound to have an air of mystery around him, and a number of feats were attributed to him that I could not verify. My favorite concerned some men who purportedly parachuted onto the top of Devils Tower to win a bet. Not thinking beyond their potential winnings, they did not have a means of descending the 865 feet to the base of this striking volcanic rock tower located near the Belle Fourche River in northeastern Wyoming. When Ginger heard of their plight, he drove to the monument with pitons and ropes and rescued them.

Ginger's ability to climb this tower and bring the men down safely did not surprise me as the first recorded ascension took place in 1893. So many people were climbing the tower that the national monument authorities had to restrict climbing so that the tower would not be permanently damaged. What did surprise me is that as the story goes, once Ginger had brought the men safely to earth, they walked away without even thanking him. Ginger apparently was not the kind of person who expected adulation, and he quietly returned to his mountain retreat in the Big Horns.

Interesting people attract attributions of feats that become legendary, and Ginger was no exception. Perhaps Trudy was one of the few people who could have navigated between the fact and the mythology of Ginger Gurrell. On the other hand, part of the fascination of Wyoming is the rich oral tradition relating man and environment. As an example, one of the Kiowa legends explaining the origin of Devils Tower claims that seven young girls were chased by bears onto a low rock. Taking pity, the gods pushed the rock into the sky where the girls can now be seen as the Pleiades in the constellation Taurus. The striations, or fluting, along the tower were caused by the bears clawing after their prey as the rock rose to the sky.

With her superb organizational talents, Trudy was involved in organizing the dude rodeos, an activity which would give today's insurance underwriters the vapors. There were barrel races and egg races in which contestants would be disqualified if they dropped an egg from the spoon they were holding while maneuvering their horses. The children would ride steers—an activity that is not as dangerous as it might sound. The steers usually ran in a straight line and bucked fairly rhythmically. There were package races in which each contestant would run one hundred yards or so to a package, undo the package, and don whatever garb was in it, usually feminine attire. The contestant then ran back to his horse, mounted, and raced to the finish line.

Although quite hilarious, the package race was a pretty tame version of the wild horse races in Buffalo, which were my favorite part of the rodeo. In the Buffalo version, a team of three cowboys would compete on the rodeo track with several other teams, each of which had a bronc on a halter. At the official's signal, the designated rider would sprint to a package, unwrap it, and don the contents, invariably underwear and a large dress. He would then sprint back to the horse. The sight of a bowlegged cowboy sprinting is unusual enough, but the sight of a rapidly approaching cowboy in drag usually caused the horses to go berserk. When the rider arrived, the next task was to saddle the horse. Many different techniques were employed. One team theorized that if a contestant chewed on the horse's ear, it would somehow calm him. I always thought it produced the opposite effect. It certainly would on me.

With the saddle screwed down, the rider would leap on board and the other team members would try to get the horse off in the right direction. Frequently, one horse would go off like a shot, circling the track in record time, but in the wrong direction, thus disqualifying the team. Many times the first horse around would have an empty saddle, which did not count either. The wild horse race was mayhem and I loved it.

As a step grandmother trying to raise children in an atmosphere that for four months of the year catered to people on vacation, Hank had some trying moments. She once felt that Jack had been rude to a guest and asked him to apologize. Feeling that he was, in fact, the aggrieved party, Jack refused. Hank confined him to quarters until he had a change of heart. Both Hank and Jack had a stubborn streak and dug in their heels. The days dragged on with Jack steadfast in his room. Sometime during the second week, Hank concluded that further confinement was pointless as the guest, who did not feel particularly injured anyway, had already departed. Jack did gain something from this experience. He discovered a bridge game in his room, mastered it, and became an excellent player, a skill that served him well in college and beyond.

While attending high school in Sheridan, Trudy and her classmates were asked to write a paper on the subject of their choice. Most students selected a topic along the lines of what they did last summer. Trudy already had a well-developed social conscience that was unusual not only for her age but for the age in which she grew up. Her topic was alcoholism, which flabbergasted her totally unprepared teacher. Hank, of course, took this choice in stride and remarked, "You certainly are in the right spot to do your research."

Guests always enjoyed the Buffalo rodeo because the ranch corral was represented by a number of contestants. Dean was a frequent team roper, and there were usually several wranglers in the calf roping events. In the 1940s and early 1950s, a seventeen-second time might win calf roping in Buffalo, which inspired many to pay the entry fee. Today if you examine your program and then take a sip of your beer, you will look up and see the calf roper returning to his horse. Bull, bareback, and saddle bronc riding were done by the younger wranglers who had not yet developed sufficient injuries and respect for the law of probabilities. Riding was also relatively inexpensive because it required minimal equipment and one did not have to own horses and haul them around.

Two young wranglers were drinking some beers one night and suggested to Jack that they all enter the bareback riding event in Buffalo. It seemed like a pretty good idea at the time. The beers were good, the rodeo was a week away, and Buffalo was thirteen miles out of sight. It all seemed kind of abstract, but the decision was made and the Rubicon, or at least Rock Creek, had been crossed. No one was about to say the next day, "You know, about last night. Well, I've been thinking and that was a really stupid idea."

The day of the rodeo came, and the three drove into Buffalo in relative silence. Contestants are given a great deal of time in which, while waiting for their particular event, they can worry and work themselves into quite a nervous state. When the waiting finally came to an end, the first wrangler climbed to the top of the chute and looked down upon a horse that stood calmly and almost idly. Yet that horse had a certain professional air about it, like a surgeon. When the gate was opened, the horse sprung out and bucked fairly rhythmically in a straight line. The rider had eight seconds to remain on board and develop a winning style. However, along about the fourth second, the bronc figured the rider was suckered and threw in a twist or two that sent him flying.

The next young wrangler climbed to the top of the chute and looked down on a monster intent on demolishing his confinement. This was the type of horse about which songs are written and stories are told around campfires. After some prodding, pushing, and tightening, the horse quieted down enough so the rider could be installed and the gate opened. With no foreplay whatsoever, this horse unwound with everything he had and in all directions at once. It was a miracle that the rider stayed on for four seconds before going flying. Unfortunately, he sustained a fairly bad fracture and the previous wrangler accompanied him to the hospital. Jack was on later so the three did not meet until the following day at the ranch. After much commiseration, one of the wranglers asked how Jack had fared.

"It went O.K.," Jack replied.

"You mean you finished the ride?"

"Actually, I won the event," said Jack without a hint of false modesty.

Like everyone, Jack had his dreams, but he also had the courage, talent, and energy required to turn them into reality. While in high school, he decided he wanted to go to Princeton as his father had. He also concluded that he probably could not get there with his current educational background, and he knew that his father would not pay the tuition of a first rate Eastern prep school. That would stop most people. Jack, however, wrote the legendary Frank Boyden, the headmaster of Deerfield Academy, and Mr. Boyden was so impressed, they struck a deal.

Jack and I were classmates at Deerfield, both graduating in 1956. At that time, most of the students had never been west of Philadelphia, if not the Hudson River. Coming from Central High School in Omaha, Nebraska, Deerfield was a major culture shock for me. Imagine what it was for a ranch kid from Saddlestring, Wyoming. Jack was involved in the usual mix of athletic and cultural activities. He was on the varsity track team where he focused on the broad and high jump. He was in the chess and music clubs and was president of the ornithology club. He always was an excellent birder and could easily identify anything that nested at or migrated through the ranch.

Living along Rock Creek in Wyoming, Jack was an excellent fly fisherman and brought those skills to the Connecticut River, which ran through Deerfield. Fly fishing was unheard of at Deerfield, and Jack was probably the only practitioner within miles. It certainly was not a topic of discussion on Park Avenue or Darien in 1956. Jack managed to land a very nice rainbow that he showed to a student on the bank before releasing it. The student was so impressed that when his mother came to visit him, he introduced her to Jack saying that he had caught a huge fish in the Connecticut River. The matron looked totally mystified and finally, in a puzzled tone of voice, said, "What in the world were you doing in the river?" It was a different world.

After a rather gloomy spring filled with torrential rains and flooding, graduation finally arrived. My parents came and were able to kill two birds with one stone as my sister, Julie, was graduating from Smith down the road to the south. My dad was beginning to show considerable signs of strain at having to be nice to so many people he did not know and would never see again when who should we bump into but Hank who was there for Jack's graduation. Hank was dressed in New England style tweeds, and except for the deep tan and signs of weathering, looked like she could be the president of the garden club in Greenwich.

Dad was overjoyed at not only seeing someone he knew but someone he liked, and said to Hank, "Let's go somewhere where we can say 'shit' and have a drink." Finding a place like that near Deerfield in 1956 was not easy, but my parents and Hank disappeared. When the three finally reappeared, Dad was in an expansive mood and much more tolerant of his fellow man, at least through the graduation ceremonies.

Jack went on to Princeton with a Navy ROTC scholarship and was quite an athlete, participating in soccer, football, crew, skiing, and lacrosse in which he was a star defenseman for three seasons. He concentrated his studies in the field of geology writing his thesis on the geology of southern Lake Valencia, Venezuela, a topic that because of its distance, was probably reasonably safe from excessive criticism. While at Princeton, he also wrote an excellent paper on the geology of the Big Horns, which is available at the ranch.

Proving that you cannot take Wyoming out of the boy, Jack attended a very formal dance while at Princeton where he caught his first glimpse of a debutante sweeping into the room in an elegant gown and long gloves. His awed reaction was, "My God. Look at the mittens on that girl."

In his senior year, Jack was awarded a Rhodes Scholarship and spent several years at Oxford. He owed the Navy a number of years of active service for his NROTC scholarship and served in Vietnam where he was a decorated navigator in the Naval Air.

In the meantime, Hank was feeling the burden of a big investment in equipment, facilities, and horses with all the attendant staffing problems and short productive seasons. Hank suggested to Jack that they sell the ranch, but Jack just said he would work until he could buy another, so fortunately the ranch was retained.

While Jack was in the service, Trudy continued to evolve into a highly interesting and complex individual with a social conscience that was nurtured in the revolutionary atmosphere of the 1960s. She became interested in Zen Buddhism and plunged into it with her usual intensity. During the fall of a year when the icy grip of winter came early to the ranch, Trudy invited a group of Buddhist monks to Saddlestring. This contingency of Asians arrived with shaved heads, flowing saffron robes, and sandals. The dudes had all departed, but the remaining staff greeted them with politeness and incredulity.

Trudy suggested a ride, and the monks were escorted to the corral where they and the horses eyed each other dubiously. Mounting was a challenge for the short Asians and the harried wranglers attending them, but everyone was finally in the saddle with stirrups adjusted. Although a short ride to the end of the lane and back would have probably provided ample thrills and been in everyone's best interests, Trudy selected Castle Rock, a ride with breathtaking vistas of the mountains and plains. Of course, breathtaking vistas naturally include a rise in elevation and exposure to the elements.

Trudy set off in the lead with the monks in tow followed by a reluctant wrangler clearly wishing he were somewhere else. As soon as they started ascending toward the ridge line of hills leading to Castle Rock, the chilly winds began blowing in earnest. By the time they had reached the top, the monks' exposed calves had turned blue, and they were so numb from the cold, they had a hard time controlling the horses. Unaccustomed to flapping saffron robes and riders exercising no authority, the horses were, quite understandably, skittish and began dancing around in the howling wind. Comfortably in the lead, Trudy was enjoying the brisk day and brilliant scenery, oblivious to the misery and chaos unfolding behind her. Although doing his best, the wrangler was essentially helpless.

The descent from Castle Rock can be intimidating under good conditions, but the passage near cliffs on psychotic horses struck terror into the hearts of the monks. They were too cold and numb to object and could not be heard in the wind even if they had tried. Trudy took their silence to mean awe and respect for the scenery and occasionally turned to smile at them benevolently while forging on.

When they finally returned to the corral, the monks were decanted from their horses, and they gazed hopefully toward the Horton House with thoughts of gallons of scalding tea. Trudy, however, had other ideas. Exhilarated from the ride, she felt another Wyoming experience would make an indelible impression upon them. She led them off and parked them on top of a wooden fence to watch the loading of cattle.

A large double-decker cattle truck was being backed into position at the loading chute. When the truck was properly positioned, the driver's door flew open and out jumped a huge, burley trucker. Glimpsing a row of teeth chattering bald Asian monks in flapping saffron robes and blue legs, the driver executed a classic double take that left him frozen in mid stride with jaws agape. In fairness, this would be an unusual sight in most of North America, much less Saddlestring, Wyoming.

When the trucker had recovered his aplomb, he walked over to Bobby Ross who was adjusting the loading gate. Bob was a fixture at the ranch for about a quarter of a century and has a subtle sense of humor and mischievous twinkling eyes. The trucker whispered to Bob, "Who the hell are those guys?"

Continuing to work on the gate, Bobby replied nonchalantly, "What guys?"

"Those guys on the fence, for chrissakes."

Bobby glanced up, continued fiddling with the gate, and said casually, "Oh, they're just a bunch of elk hunters."

The sight was a show stopper for more than just the trucker. Dean, who lived in a cabin ironically named Virtue's Reward and situated on the hill above Horton House, was returning home when he heard tinkling bells and weird chanting emanating from the house. Pondering the ethics of the situation for approximately five seconds, he said he hid in some bushes and peeked into the window. After all, this was a sight that would fuel stories on a cold winter's night for decades to come.

After being released from the Navy, Jack spent time at the ranch and in Washington where he worked for Walter J. Hickel, secretary of Interior in President Nixon's administration. Hickel visited the ranch as Jack's guest in the early 1970s. Washington did not seem to be Jack's natural milieu, and during the 1970s, he spent more and more of his time at the ranch or in his natural resources consulting business.

Jack and I went our separate ways after Deerfield. I went west to Stanford and then spent two years in the Army in France. Fortunately, it was too early for Vietnam. After the service, I began a banking career on Wall Street and lived on Manhattan's upper East Side. During one excruciatingly hot and crowded ride on the Lexington Avenue subway, I began to think of the ranch and backpacking in the coolness of the high country. With each subway ride, the idea began growing into an obsession, and even my wife, Beth, who is from Montclair, New Jersey, and had never been backpacking in her life, began showing signs of enthusiasm.

I wrote Jack inquiring about a stay at the ranch and then using the cabin at Willow Park as a jumping off point for a backpacking adventure. He replied as follows:

> If I were you, I would adjust total time in Wyoming to allow more time in the mountains and less in the low country. This is a personal (and obviously not a business) opinion. The country west of Willow Park, in the wilderness area and at timberline, is superb. One idea is to take off backpacking from the cabins and work timberline around to Spear Lake. Or better, from the cabins via back country to the old reservoir, up the tarns to Gem Lake,

and then across timberline to Spear. This is where the elks calve and summer. All the creeks start here and the topography, as you know, is glacial. There are even a few mountain sheep. Except for the possibility of rain, the twenty-first of June is a good time. The season hasn't begun and you should have the country to yourself. Regretful am I that I wouldn't be able to play even a small part.

Jack wrote this February 15, 1967, on ranch stationery, the design of which has remained unchanged for more than thirty years. He described the trip as if they were casual day outings, and for an athlete of his caliber, they might be. I have backpacked most of the terrain he described, and some of it is enough to make a banker sit down and whimper. As it turned out, snow was more of a problem than rain on that particular June day, and some of the creeks were impassable for a backpacker. Nevertheless, we had a delightful time and totally forgot the Lexington Avenue subway.

The high country casts a spell that can change a person, at least temporarily. Jack was a different person in the mountains. He was more relaxed, carefree, and philosophical, and I think he was basically a highlander at heart.

Although born and raised in Wyoming, Jack had acquired some tastes in his travels that truly mystified some of his compatriots. He decided to climb Mt. McKinley in the Denali National Park in Alaska which, at 20,320 feet, is the highest mountain in North America. Reaching the summit, as Jack eventually did, requires incredible physical conditioning and stamina along with luck because bad weather frequently forecloses the last leg of the trip.

To condition himself for the trip, Jack began a regime whereby every day he would don hiking boots and shorts, fill a backpack with rocks, and walk up to the top of the Speedway and back. The climb is steep and takes thirty to forty-five minutes. Although vigorous exercise, it is still short of the McKinley summit by about 14,000 feet. Nothing can totally prepare a person for a trip like McKinley, so basic endurance and courage are required.

Jack realized that most of the ranch hands would think he was nuts, and he particularly wanted to avoid Dean. Like most cowboys, Dean was not lazy but could be considered economical in terms of energy expended. To Dean, energy was for work, and the idea of exhausting oneself for pleasure was nonsensical. Consequently, Jack went out of his way to undertake his exercise when Dean was not around.

Of course, the inevitable happened. Jack was rounding the corner of the barn in his hiking attire and ran almost literally into Dean, who was coming from the opposite direction. Dean eyed Jack's clothing skeptically and asked, "What do you have in the pack?"

Jack replied somewhat sheepishly, "Rocks."

"Where are you going to take them?"

"Up to the Speedway."

"What are you going to do with them once you get up there?"

"Bring them back down."

Dean walked away incomprehensibly muttering to himself and shaking his head. Jack knew that explanations would be futile.

One day, Jack was riding near the north fork on the Sawmill Trail when he saw a timber wolf, an incredibly rare sighting. Environmental preservation was in its infancy and considered rather subversive. Ranchers had pretty much extirpated anything remotely carnivorous, so Jack watched the wolf in awe as it disappeared into the pines. He then rode to some exposed ground and examined the tracks in the mud.

On returning to the corral, Jack saw Dean and remarked enthusiastically that he had just seen a wolf. Dean gave him that sad look reserved for a retarded child and said that it must have been a coyote. Jack knew better than to argue but, later in the week, suggested to Dean that they check a fence on the Sawmill. He led Dean by the wolf tracks and heard a very satisfactory, "My God! Hold up."

Jack stopped his horse and asked what was the matter. Dean had dismounted and was squatting by the tracks examining them intently. He finally said, "These are wolf tracks."

Jack looked at them quizzically and replied, "Naw, they're probably coyote."

Dean gave Jack the slow learner look and said in a professorial tone, "Note the size of this track. It's about two inches bigger than a coyote's. Now take a look at the pad on the rear paw. This one's the rear paw. See how the pad is triangular and lobed. A coyote's is different. It's a wolf all right."

"Well, Dean," Jack said appreciatively, "You can sure spot 'em."

One of the more enduring stories about Jack concerns a fisherman who was crawling toward Rock Creek. The guest was crawling because the best fishing is usually the most inaccessible. To find a good spot, anglers look for a location where they must cross a barbed-wire fence, slither down a virtual cliff, walk through a bog, and then crawl through willow thickets to reach the stream. All anglers are prepared to lose their fly into overhanging brush on the first cast. It is truly a great sport.

The guest was on hands and knees, concentrating intently on guiding the rod tip through the brush, when he almost literally landed on top of Jack and a young lady *in flagrante delicto*. The guest, a cherubic looking banker from the West Coast with courtly manners, doffed his cap, murmured something about the weather and fishing further downstream, and crawled a slow, but deliberate, retreat to the rear.

Years later, I happened to meet the guest, and we enjoyed a lot of good laughs over Wyoming stories. When the tale of Jack and the stream Siren came up, the interloper said that they were not exactly *in flagrante*. I stopped him there, not wanting to know because the story had acquired its own reality and mental image.

Trout fishing along a Wyoming stream is more of a metaphysical exercise than an application of artificial lure to water. The slow, graceful rhythm of the fisherman blends into the surroundings and amazing things happen when you become integrated with nature. I was once caught in a sudden, but brief, downpour without rain gear. I crawled into a thicket and cut some broad rhubarb leaves to cover my head, shoulders, and knees. The hastily improvised poncho must have been effective camouflage because a doe backed into the thicket and waited out the storm within arm's length of me. I can still picture the rain dripping off her nose.

Another fisherman had fallen into that blissful hypnotic state that comes with a series of perfectly delivered casts in a beautiful environment, when the sharp crack of a high caliber weapon immediately behind him shattered the trance. For a split second, he was transported back to Korea in the early 1950s. Turning, he saw a V-shaped ripple in the water formed by a small nose disappearing into a bank den. The warning slap of the beaver's tail had produced a heart stopping scare, the intensity of which could only be known by the fisherman and his laundry service.

Fisherman have unforgettable encounters with nature, stumbling on to small fawns exploding out of tall grass or wild turkeys maneuvering like fighter pilots through the trees. One respectable, and reasonably sober, guest claims to have seen a bear, and another, a mountain lion. I suspect large dogs, but who knows? That is the mystery of the stream.

What happened between Jack and the young lady on the stream bank that beautiful summer day in Wyoming? The rational, and perhaps, cynical mind might say the rocks, brambles, and mosquitoes of a creek hardly make an ideal Cupid's bower. The more romantically inclined might conclude that the warmth of the day, the song of the lazuli bunting and the yellow warbler, and the murmuring waters might be the perfect catalysts for a little *delicto* between warm blooded youth and a stunning stream nymph. Who knows or cares? The story now has a life of its own. The rest of the stories, however, did happen—pretty much.

On Jack's death in 1981, Malcolm Wallop, a neighboring rancher, friend, and U.S. senator, read a long and moving tribute into the *Congressional Record*. He said in part:

> Jack Horton turned the opportunities of his life into something straight out of a Michener novel. He lent color to ordinary experiences. He lightened an exceptional academic background with flashes of humor and the absurd. He cajoled and encouraged his professional colleagues toward unexpected achievement. He invented language so that western resource users could learn to talk to western environmentalists. And in the midst of this perpetual professional motion, he played a killing game of tennis, rode the rodeo circuit, climbed mountains, and ran the Boston Marathon.

The story of Jack and Trudy is brief and one of unfolding promise and potential that remains unfulfilled. Star-crossed genes caused both to die of cancer, Trudy at age 29 and Jack at 43. Fortunately, Jack before his death met Margi Schroth, who saved and preserved the traditions of the ranch.

Frank "Skipper" Horton with his first wife, Gertrude S. Butler, parents of Bill, Jack Sr., and Bob. Courtesy of Pam and Mike Dixon, San Rafael, California.

Looking west from the Castle Rock ridge line, the snow-covered alpine region of the Big Horns is in the background. In the foreground, Stone Mountain is to the left, the jagged rocks outline Red Canyon, and the Speedway protrudes into the foothills. Photo by Beth Morsman, Deephaven, Minnesota.

Henrietta "Hank" Horton, the postmistress of Saddlestring, Wyoming, sorting the mail. Courtesy of Shirley Taylor, Kirby, Montana.

Skipper displaying his natural exuberance. Courtesy of Pam and Mike Dixon, San Rafael, California.

Dean in the rodeo field in 1967. Courtesy of Flora Westin, New York, New York.

Dean at the corral. Courtesy of Pam and Mike Dixon, San Rafael, California.

The author and Dean making a wreck of the "Ole 97." Photo by Beth Morsman, Deephaven, Minnesota.

The teenage Trudy Horton with the Castle Rock ridge line in the background. Courtesy of Margi Schroth, Saddlestring, Wyoming.

Jack Horton in his natural element in 1979. Courtesy of Elinor and Peter Constable, Washington, D.C.

Jack in naval dress blues toasting the Taylor family, fellow ranchers, and friends, 1966. Courtesy of Susanna Taylor Meyer, Woodbury, Minnesota.

Trudy in November 1966 with daughter, Anne, and son, Will. Courtesy of Anne Dixon, Longmont, Colorado.

Jack as assistant secretary of the Interior for Land and Water Resources. Courtesy of Margi Schroth, Saddlestring, Wyoming.

Margi, Dean, and Hank on the ranch house lawn. Courtesy of HF Bar Guest Ranch, Saddlestring, Wyoming.

Margi Schroth, owner and manager. Courtesy of HF Bar Guest Ranch, Saddlestring, Wyoming.

(HELP!)

The man standing before the judge gazed dully at the floor through bloodshot eyes. Disheveled and unkempt, he seemed quite indifferent to the judicial proceedings that were deciding his fate and residency for the near future. Having been caught red handed in an attempt at petty larceny, the defendant had not given the public defender much of a case.

Consequently, the lawyer, who was a conscientious but very busy public servant, had decided on a plea bargain. Because there was little with which to bargain, he and his client needed a miracle, and as luck would have it, one marched purposefully into his office in the form of Hank Horton. Why would a person of Hank's power and prestige in the county espouse the cause of a petty felon? Well, this particular thief happened to be a passably good cook at the ranch, and it was the height of the season.

Knowing the odds against replacing a cook on short notice, Hank was understandably desperate and felt that a gentle hand on the scales of justice might prove beneficial to everyone. She argued reasonably that a jail sentence would merely cost the frugal taxpayers of Wyoming room and board. As the man was more stupid than dangerous, Hank proposed a probationary period of three months during which time the defendant would continue his cooking duties at the ranch, fines and court costs would be withheld from salary, and once a week, he would check in with the court. A true win/win situation for the court, taxpayers, defendant, and the hungry ranch guests.

Delighted with this unexpected bit of luck, the public defender discussed the proposal with the judge, another very busy public servant, who endorsed it enthusiastically. A hearing was immediately called, and the judge, defender, and Hank looked benignly at the defendant, each thanking the stars for a positive solution to a sticky problem.

The judge explained to the defendant that he had the choice of a six-month sentence at the correctional facility in Douglas or a three-month probationary period during which he would continue to cook at the ranch. The defendant's brow furrowed and he appeared to fall into a deep concentration. As the wheels in the man's brain slowly turned, the judge, public defender, and Hank stared at him with growing incredulity. This was not a difficult decision.

In disbelief, the judge asked the defendant if he understood the choice. He nodded affirmatively, and the judge said, "Well, what is your answer?"

Finally, the prisoner said, "I'll take Douglas."

Finding and hiring adequate help was a significant and painful challenge for Hank, and one of the issues she raised when she discussed selling the ranch with Jack. Wyoming has a small population, and the work at the ranch is highly seasonal. Like many western states, Wyoming has seen its share of boom-and-bust economies precipitated by such factors as cattle and energy prices. These cycles affect the availability of help along with unusual occurrences with unintended consequences. For example, when the Powder River Basin coal mines opened with well-paying jobs and benefits, Hank and others had quite a time staffing, particularly in the more menial jobs. As a result, her search for help went far beyond the borders of Wyoming.

In the staffing arena, the corral has always been an island of relative stability compared with other areas, particularly the kitchen. As most guests come to a dude ranch to ride, the corral is a focal point, and fortunately, the HF Bar corral has always had a core of highly competent and cheerful professionals complemented by several itinerant wranglers.

Dean was, of course, part of that core before he became foreman. He worked with some truly amazing characters whose lives bridged the turn of the century and whose experiences defined what it meant to be a cowboy in the developing and evolving West.

Barney McClain was a fixture in the corral from the 1930s through the 1950s. He was born around 1890 in the Oklahoma Territory and, as a teenager, served some time for cattle rustling. He was never particularly ashamed of that period of his life, as many of today's respectable ranch families got their start in that era with a "long rope." During World War I, Barney served in the mule troops and lost a forefinger in a dallying accident. Wrapping a rope around a saddle horn can be tricky business when there is one thousand pounds plus of unpredictability on the other end. It was a lesson Dean never forgot, and he often cautioned younger wranglers to watch their dallies.

Barney, who could neither read nor write, was married to a high school teacher in Sheridan. He wore a tall crowned gray hat that may have been white in the beginning and had a perpetual hand-rolled cigarette dangling from the corner of his mouth. His shirts were always solid colors, usually white, and they were buttoned at the throat to keep out the dust. The old timers considered snap shirts for sissies and never wore them. I suspect Dean did not switch over until after Barney was gone.

The guests had a great deal of respect for Barney even though he could never seem to remember their names. Cowboys have amazing memories for horses and other livestock. Although they may forget the name of individuals, they can always remember the horse each rode even over the course of decades. Barney frequently called dudes by the name of the horses they rode.

Assigning and remembering the names of horses had a practical aspect when you had to select a particular horse out of one hundred. The wranglers named the horses, and in the early years, the names were not always politically correct. Of course, the ranch hands had not been to any sensitivity seminars. The names tended to fall into categories, such as place names, for example, Manville and K.C. ; colors, for example, Carrot and Spade; drinking, for example, Oley (for Olympia beer) and Jigger; physical attributes, for example, Altitude; and children's horses, for example, Garters and Chuckles. Barney, of course, named many of the horses, and the appellations ranged from hilarious, to ironic, and occasionally, inscrutable. His own horse was called Potty.

Barney had a great sense of humor and loved practical jokes, especially his ritual for initiating new wranglers. His initiation rite depended on a gelding named Lipstick. Lipstick was considered the perfect horse for the ladies because he was gentle, patient, and reliable—except for the first ride of the year. For some unknown reason, Lipstick would stand calmly for that first saddling and then go absolutely berserk when mounted. This one incredible bucking tour de force seemed to exorcise whatever devils plagued him, and Lipstick was perfectly benign for the balance of the year.

Almost forty years later, a former wrangler described the experience as if it were yesterday. He had just arrived at the corral when Barney, pointing to Lipstick, said, "Saddle up that women's horse and let's get to work." Barney explained that Lipstick was a good wrangling horse and that because of his gentleness, a new wrangler could learn the ropes without difficulty. It was only later that the initiate realized the significance of the unusual number of people hanging around trying to look busy while watching surreptitiously.

The new wrangler swung into the saddle and looked to Barney for directions. Barney looked back with a grin spreading from ear to ear. There was a pregnant pause of approximately two seconds before Lipstick came alive and began to unwind. The horse went straight up and then began a series of fishtails, each of which left a larger and larger gap between the rider and the saddle. The young wrangler grabbed for the horn, which seemed to be growing further and further out of reach. One powerful lurch sent him sailing over the horn landing with his legs wrapped around Lipstick's neck. Horses are unaccustomed to riders wrapped around their necks, and Lipstick pause momentarily. He then gave a mighty shrug and jerk and sent the wrangler sailing.

It seems to be a rule of horsemanship that when you are thrown, you land in manure. Wyoming covers almost one-hundred thousand square miles, and it cannot all be covered with horseshit. Nevertheless, it is a good bet that you will find some when you are thrown. Our young wrangler did—a big pile of it. Barney was helpless with laughter as were the other onlookers. Barney never tired of this initiation ceremony and, with Lipstick's help, was able to pull it off for many years. I suspect each initiate was pledged to secrecy.

A little bit of Barney and Dean rubbed off on each other. They were both magnets to kids and returned the respect shown them. Both could often be seen walking their horse up and down the lane with a gleeful and proud kid in the saddle. Barney and Dean were also team roping partners, Dean being the heeler. They practiced frequently in the evenings and did reasonably well in the local rodeos.

Like many cowboys in the 1930s, Jack Buckingham and his wife, Babe, went where the scarce wages were in those depression years. As cowboying jobs were hard to find on the financially troubled ranches, Jack worked in a lumber camp where Babe did the cooking. When an opening occurred at the HF Bar, Jack went to work at the corral and Babe took charge of the cabin girls.

Bob Ross came to the ranch as a guest and stayed on about a quarter of a century. He was, and is, a fine drummer who used to play with a guest who had set up an organ in the cabin, Misty Moon, and another guest who was a credible chanteuse. He also played occasionally at the ranch dances, which were heavily oriented to square dances, waltzes, and the "Varsouviana" or "Put Your Little Foot." The dance fare is much more eclectic and modern today.

Barney, Jack, Bob, Dean, and a few others over the years made up the nucleus of the corral. This nucleus conducted on-the-job training and made sure that none of the other wranglers ever became overconfident or too impressed with his own self importance. One former wrangler recollected,

Late in the fall of 1957, before I left the ranch to join the Army, another wrangler and I took an afternoon off to go deer hunting near Spring Creek. We found a good concealed position among some rocks and waited. In time, several deer appeared, and of course, the other wrangler got one with his first shot. I shot and missed, and shot and missed, and missed, and missed. I had borrowed a .25/.35 lever action saddle gun that held about a dozen rounds. It was like a really bad western movie. There I was, now standing up, blazing away as if at some bad guy. It was quite by accident that I managed to kill the deer. By then, it was late in the afternoon, and we hurried to get back to the ranch to dress the carcasses, so we could get to supper.

In those days, supper in the Help dining room was served at two long tables. Fairly bursting with pride as this was my first deer, I took my place and waited for the inevitable questions. None came, but I and everyone else overheard Dean and Barney, who sat together at the other table, talking *sotto voce*:

Barney: "Was right noisy out there today."

Dean: "Yep. Could've been the Battle of Little Big Horn all over again." That's all there was, other than, perhaps, the briefest snort from one of the old irrigators.

The most junior wrangler was often assigned to take the children out for rides, a task that was easily a first step toward sainthood. The children made up the "mosquito fleet" and countless hapless, but cheerful, wranglers were assigned to lead them off at an excruciatingly slow pace to prosaic spots transformed with hyperbolic names such as Bear Belly Gulch and Buffalo Cave. The kids' horses are marvelous —small, docile, good-natured, and patient. My father swore that if I started to list, often from falling asleep, Garters would sidestep back underneath me.

Once, a group of parents and a new wrangler took off with a bunch of kids for an overnight beyond Stone Mountain at Ginger's cabin. Given the extended time it takes to organize children and their related paraphernalia, a late start was inevitable, and the sun naturally got pretty hot as the group zigzagged slowly up the mountain. By sundown, everyone was ready to call it a day. The horses were hobbled, the campfire extinguished, and everyone went to bed. Normally, hobbling the forefeet will keep horses within easy range. And because horses are fairly gregarious, they will normally stay in a group. As an added precaution, you can put a bell on the more free-spirited of the horses to make them easy to find the next day. This procedure works well most of the time. Jack Horton used to swear that the horse that ranged the furthest always had a guilty look. I could not tell the difference.

The next morning, the horses were gathered fairly easily, but a quick count revealed that one was missing. The stray animal was Mouse, a horse beloved by generations of children and never considered particularly adventuresome.

Everyone began a thorough search, first on foot and then on horseback in widening concentric circles. As the hours wore on, the search became more and more intense and far ranging to the point that the wrangler feared that a guest might become lost in the pine forest. People were not only worried about Mouse, but honor was also at stake. Losing a horse is a rare and ignominious event. Losing a hobbled horse is one for the record books. Although this mishap could have happened to anyone, the guests were delighted to be accompanied by a wrangler so that the blame could be placed squarely on a professional.

Finally, the search was called off, and the party returned with Mouse's rider doubling up with another. The corral staff was incredulous, and the poor wrangler suffered ample humiliation, although no one said a word. For several days, wranglers went searching for Mouse but always returned empty handed.

Six days after Mouse's disappearance, as the cowboys rose and stumbled sleepily down the steps of Wrangler's Roost, they found Mouse standing patiently on the other side of the gate to the corral still wearing hobbles and only a little worse for wear and tear. What happened remains a mystery. The route back from Ginger's cabin is fenced and there are two gates to cross. The only way back without fences is by way of the aptly named Lost Trail, which begins where the middle fork of Rock Creek joins the south fork and wanders through bogs and fallen timber to the Balm of Gilead behind the Speedway. To avoid gates, Mouse must have gone down the south fork canyon through six tough fords. Whatever the route, accomplishing it in hobbles was an incredible feat. Perhaps Mouse woke in the middle of the night and figured he should be at home. Who knows? I guess all you can say is, "Well done, thou good and faithful servant."

As good as Dean and other members of the corral nucleus were, they could not prepare a young wrangler for every eventuality, particularly those involving the idiosyncrasies of guests. A wrangler once took a group of dudes over Stone Mountain for a picnic. As the day was hot, the libations started flowing freely, and one woman became rather tipsy. The return to the ranch was long and hot, and the descent of Stone Mountain can be jarring. At the top of the mountain, the wrangler solicitously inquired if anyone would care to take a rest before descending. For most of his life, he had said "to take a leak," but the corral had taught him the appropriate euphemism for use with guests. As everyone was eager to return home, no one took advantage of the opportunity.

About halfway down Stone Mountain, the tipsy lady let out a shriek and burst into hysterical laughter. The wrangler, who was in the lead, stopped, turned back to the woman, and politely inquired if anything was wrong. He had a rather bad feeling about his inquiry and somehow knew he was going to get the worst part of the exchange. Nevertheless, he felt he could not ignore a shrieking guest.

Between guffaws, the lady announced loudly and gleefully, "I have just wet my saddle." On reflection, there is absolutely no satisfactory response to this type of statement. In fact, there was nothing at all that could be done. The twenty-year-old wrangler had not yet learned that the only response is a polite chuckle, a shrug of the shoulder, and then a return to the trail at a much faster pace.

The wrangler thought he should say something, and as his mind furiously considered the meager alternatives, he firmly wished he were somewhere else. Finally, he mumbled, "Are you all right?" The inanity of this response set the woman off into further gales of laughter. Although only an hour from the ranch, the poor wrangler endured the longest ride of his life.

In contrast to the corral, the kitchen always provided Hank with more than enough challenges, distractions, and exasperation. Feeding between one hundred and two hundred demanding guests and staff can be difficult under the best of situations. Orchestrating the convergence of reliable and experienced food preparers, a thinly stretched comestible distribution system, servers with short attention spans, and guests who think they are still on Park Avenue into a glorious culinary experience at the end of a gravel road thirteen miles west of Buffalo, Wyoming, is a task of herculean proportions. That it works so well is a tribute to all concerned. Occasionally, a touch of patience and a sense of humor can be helpful.

In the 1950s, many of the chefs and bakers came from Las Vegas, which was somewhat off season when the ranch was in full swing. One baker normally worked at the Desert Inn and considered the ranch a kind of vacation. He recruited chefs and bakers from other desert resorts, and they did a superb job while having a great deal of fun at the same time. They lived in either the ranch house or a cabin ironically called Cupid's Reward, and they tended to drink prodigiously, fortunately after dinner.

One evening, two Filipino chefs shattered all previous alcohol consumption records and got into a significant altercation. One chef excused himself and reappeared shortly with a rather menacing butcher's knife. Astonished guests watched in horror as the armed chef chased the other, both screaming at the top of their lungs in some obscure Tagalog dialect. This is an unusual occurrence in the Big Horns. Even normally hardened New Yorkers were nonplused. Fortunately, the two chefs were overweight and in terrible shape so their running became lurching in less than a minute. A burly wrangler was able to grab the chef with the knife from behind, and he happily dropped the knife as he was totally exhausted by now and in a pacific stupor. The next morning, the two chefs returned to being the best of friends and seemed to have no recollection of the incident.

When the Las Vegas connection finally ended because of retirements and diminishing seasonality in the gambling mecca, Hank had more problems in retaining reliable chefs. The difficulties were frequently as much personal as culinary. One cook ran away with the salad girl. Such an event is normally not disastrous as such a pair can usually be persuaded to return when they sober up and reality begins to set in. In this case, the particular salad girl happened to be married to a candidate for a sheriff's office in another state. Although the cook was good, Hank decided that plenty of distance between all parties was in the best interests of everyone.

Sometimes tradition gets in the way of reality. For years, the ranch had a Sunday night buffet that featured smoked salmon and what Hank called, "Acres of Jell-O." Getting a smoked salmon to the ranch on a weekly basis became a tiresome ordeal. Because of the tradition, guests expected the salmon, and it was generally exhausted before the arrival of those who pushed the cocktail hour to its logical and tardy conclusion. These guests would gaze balefully at the carcass of the salmon and then have no choice but to graze amidst the acres of Jell-O.

One inexperienced cook had never seen a thirty-inch smoked salmon before arriving at the ranch. In the Big Horns, the trout are generally of much smaller proportions. When asked to get it ready, he eyed it dubiously and then grilled it at high temperature. After all, cooks are paid to cook. The smoked salmon tradition died off quietly shortly thereafter.

The dishwashers were a particularly mobile group, constantly responding to greater opportunity or remaining one step ahead of the law. Hank soon gave up asking for their résumés when she discovered that in many cases, rap sheets would have been more appropriate. Dishes must be washed, and Hank went to great lengths to keep the jobs filled. As an example, she once gave the following list of errands to an innocent, young employee she sent into Sheridan:

Pick up 2 bridles, 1 hackamore, and 3 saddle blankets at Kings
Charge 2 large sacks of flour
Pick up and charge laundry
Drop off guests' film
Bail out and return dishwasher with attached cash

A busy and occasionally understaffed kitchen can be a highly stressful environment, and the customer is not always right. One exceedingly demanding matron was constantly returning food for minor adjustments. When breakfast was brought to her one morning, she complained that the cantaloupe was too cold and should be warmed in the sun or on top of the oven. Also, the bacon could stand a little more crisping. The server returned the items to the cook and relayed the instructions. Sensing that the harried cook had reached the flash point, the server instinctively stepped to the side just as the cook flung his spatula at the door to the dining room, which another server happened to be opening. The spatula clattered into the dining room followed by an extremely audible, but impractical, suggestion about what the matron should do to her person. Most of the woman's fellow diners not only felt the outburst was justified but also concurred with the suggestion. Normal conviviality resumed immediately.

The servers have always been called hashers, a decades-old term that belies the quality of the food. Hashers typically have been, and are, prep school or college students who are the children of guests and have been guests themselves. While solving a particular labor problem in tight markets, this choice of employee presents its own unique management issues. While generally cheerful and reasonably efficient, the young hashers tend to be hyperactive, free spirited, daring, unmotivated by the usual economic considerations, and highly hormonal. For years, the male hashers lived in a small cabin aptly named the Passion Pit that served as a social center for gathering, drinking, singing, and other activities that when discovered, tend to shorten a parent's life expectancy.

One bright July morning, a young hasher struggled into his uniform of blue jeans, checkered shirt, garish and outrageously clashing tie, and white linen jacket with just under the maximum tolerated stains. Forty years ago, few people had cars with which to travel to town for entertainment or weekends and days off. Consequently, most of the staff stayed on the ranch all the time, and weekends were somewhat indistinguishable from weekdays. In fact, for the youth, most every day was a weekend.

The night before was no exception for this hasher who awoke that morning with a splitting headache and other accouterments of a first-class hangover. With downcast eyes, he walked by the creek on his way to breakfast at the ranch house and happened to notice a frog sitting on the bank side. Absentmindedly, he picked up the frog and put it in his jacket pocket. His befuddled mind had not contemplated a use for a frog. On the other hand, one never knows when a frog might come in handy.

In his present state, the hasher did not look forward to his first customer who was always pacing in the ante room until the bell rang at 7:30 a.m. As the clapper struck the first note, the guest strode imperiously to his seat and demanded the usual—a three-minute boiled egg. Now a three-minute boiled egg may be delicious in Cincinnati, but at fifty-five hundred feet, a three-minute egg is bilious goo. The routine was the same every morning: When first presented with his boiled egg, the guest removed it from the egg holder which he then turned over to reveal the larger receptacle at the other end. With a critical eye and the precision of a surgeon, he cracked the egg and poured the contents into the receptacle. Glancing at the result, he recoiled in horror and pushed the egg away. One would think a doctor would have a stronger stomach for such a sight. The egg was returned to the kitchen and the process repeated several times until it was declared edible. The overworked cook, of course, seethed and nearly had to be restrained every morning.

On this particular morning, the guest returned the egg twice and our hungover hasher incurred the wrath of both guest and cook. Feeling like an ill-treated ping pong ball, the hasher waited sullenly while the new egg boiled. Thrusting his hands into his pockets, he unwittingly discovered the means for revenge. Thinking there must be a just God, he carefully placed the squirming frog into the larger receptacle of the egg holder with the newly boiled egg in the upper cup.

Returning to the guest's table, he hovered solicitously while the doctor removed the egg, turned over the cup, and went into orbit. Even after forty years, the event has not lost its luster for the hasher who still glows contentedly while recounting it. The effect was exceptionally gratifying, if not spectacular, and well worth the risk. When some semblance of normal speech finally returned to the guest, he sputtered, "I'm going to report you to Mrs. Horton." Drawing himself to a position of attention and paraphrasing that marvelous line from *Gone with the Wind*, the hasher replied, "Frankly, Sir, I don't give a shit," and withdrew to the kitchen. Oddly, the guest and the hasher got along just fine afterwards, the incident was never mentioned again, and for the rest of the guest's stay, the first egg was always satisfactory.

Before the era of efficient food distribution, the ranch was almost self-sufficient in the production of comestibles. Vegetables, beef, hogs, and chickens were processed on the spot and provided hashers with the occasional unpleasant duty. By far the worst was the weekly fried chicken lunch on Sundays. Two hashers were required for the executions, and this duty was rotated with enough infrequency that it always came as an unpleasant surprise. The chickens gave the hashers a merry chase, and the hashers always arrived at the kitchen door covered with blood and feathers. No one could have been happier when food distribution replaced self sufficiency.

Hank's relationship with the hashers was caring, yet pragmatic. She had no desire, nor ability, to act *in loco parentis*, yet needed to apply enough discipline to ensure satisfactory service. One warm, dark August evening, the hashers and the cabin girls were skinny dipping in the pool and making a lot of noise doing it. Such is youth. When one is older and naked, one tends to be quiet about it. Looking toward the Horton House, they saw a flashlight bobbing its way toward the pool. Instantly, naked coeds leapt from the pool and scattered into the trees. Silence returned, which is all that Hank wanted. The help probably construed her intervention as a moral warning.

At one point, Hank became so upset with the noise level in the ranch house that she put a cot in the living room and spent several nights there. Although this solution worked well, Hank concluded she could not go through life sleeping on a cot just to keep things quiet. Noise did bother her though. She once stood in the entry door to the kitchen to try to determine the source of the noise. Unfortunately, a new hasher charged through the wrong door and decked her. She picked herself up and wandered off muttering.

The hashers generally ate from a different menu than the guests and needless to say, it was considerably more Spartan. Nevertheless, they were always ingenious enough to feed themselves well. One morning, the breakfast cook was puzzled to encounter so many orders at a very early hour. He had filled at least six orders, yet when he peeked into the dining room, there was no one there. Opening the door to the pantry, he uncovered three hashers crunched together guiltily stuffing themselves.

Hashers could always amuse themselves when the work became drudgery. As an example, they established pots for virtually any purpose such as for the person who dropped the next tray or wore the worst tie. As acute as their sense of humor was, they could not always decipher Hank's wit. Some appeared quite puzzled when Hank announced in an exasperated tone, "This linen closet looks like a pet shop window."

Although most structures at the ranch are predominately wooden, there have been surprisingly few fires over the years. For this reason, the conflagration of the Passion Pit was a memorable event. It went up in flames some time after dinner, and no one ever learned the cause. The usual suspects of faulty wiring, cigarettes, etc., were enumerated, but the destruction was total, so we will never know. Fortunately, the Pit was completely unoccupied as the hashers were cleaning up after dinner. The bad news was that the fire got a pretty good start before it was noticed.

The dinner bell was rung, and as everyone had just eaten, they could either assume there were seconds on dessert or something was amiss. The latter choice had by far the best odds, so guests and staff gathered at the scene, while someone called the Buffalo fire department. Flames and smoke were rolling out the windows and approaching the building would be foolhardy. The spectators could feel the intense heat on their faces. Suddenly, a youth in a white linen jacket broke through the crowd and ran toward the building. The crowd gasped and pleaded with him to stop, but he paused only briefly at the door for a lung full of air and then plunged into the inferno.

What sentiment could possibly draw a young man into such danger? A letter from a loved one? A locket with a wisp of tresses? A picture of his mother? Moments later the man staggered through the door enshrouded in smoke with tongues of flame licking at his legs.

He was dragging his mattress. His mattress? One can only guess what prodigious feats were accomplished on this simple piece of bedding. One can only hope that after a quarter of a century, the man has this mattress enshrined somewhere so that he can gaze contentedly at it and regain strength when the spirits and flesh are flagging.

A brief moment of hope arose as Dean came struggling down from the barn unrolling a fire hose. When he reached the Pit, he signaled to a wrangler to turn on the spigot. The expectant crowd watched as a few desultory spurts emerged from the hose followed by some gurgling and hissing, and then about a pint of water poured out on Dean's boots.

By the time the fire truck arrived, the Pit was reduced to smoldering embers. In defense of the fire fighters, it is a long trip and the building was consumed faster than Gomorrah. One fortunate hasher had just worn the last of his clean clothes and had put them in the coin-operated laundry called the Palace of Democracy located behind the kitchen. He had deferred this chore for as long as possible, and this creative bit of procrastination meant that he was the only one with clothes. Hank contributed some blue jeans to the smaller ones, but even the shortest appeared to be wearing pedal pushers. Within a few days, life was back to normal, but the memory of the legendary mattress remains.

Just as fires on the ranch have been rare occurrences, fires in the nearby forest have fortunately been equally rare, small, and well contained. Sixty years ago, a bad fire in the Circle Park area miles to the south produced acres of fallen timber that only now are regaining signs of reforestation. A number of years ago and late in the season when the guests were down to two tables in the dining room, a man announced that he had seen smoke behind the Speedway. As he happened to be under the influence of a large dose of cocktails, no one believed him. The next morning, a very sober postman announced the same observation, which, this time, had instant credibility.

There was, indeed, a fire and a team from the forest service arrived. Surveying the scene, they decided to tackle the problem on foot so they would not have to deal with panicky horses if the fire were severe. This was a mistake. The shortest way to get behind the Speedway into the Balm of Gilead drainage is over a trail called the Martini. The trail was obviously named by someone with a substance abuse problem. With a great deal of imagination, innate or induced, you might construe a cup-like rock formation near the top as a martini glass. Whoever named the Martini Trail, also named a trail that joins it the Olive. I assume this person was a guest, as very few Wyoming cowboys consider the martini their drink of choice.

The few remaining guests all pitched in and made sandwiches and lemonade for the valiant crew, which took off loaded down with equipment and refreshments. On a clear day, as this one was, the trail looks deceptively short and not particularly elevated. Because the trail has no views and is basically uninteresting, few people travel it. Consequently, there was no one handy to warn of its rigors. And rigors they are. The trail goes straight up with no elevation attenuating switch-backs. As it is infrequently traveled, it tends to be covered with fallen timber and rough spots from erosion.

Towards the end of the afternoon, the crew came straggling back to be greeted by expectant guests and staff. Covered with dust turned to rivulets by sweat, the crew showed definite signs of exhaustion. Asked if the fire had been extinguished, a spokesman replied ignominiously, "Not exactly." Further questioning revealed that the fire fighters were not in good enough shape to get to the top of the mountain, much less down the other side. In fairness, it is a grueling hike. Fortunately, the fire died out naturally before another crew had to mount an assault.

The abilities of the cabin girls are usually overlooked. These girls reestablish order in the chaotic condition of the guests' residences. They seem to deal cheerfully with the mud, litter, and debris that is natural in a country environment. Decades ago, the cabin girls were generally local, but better employment opportunities forced Hank to rely on the daughters of guests who came for the western experience and an active social life.

Years ago, a cabin girl who was in reality a somewhat older cabin woman, had the habit of tippling on the guests' ample supplies of opened and available booze. By the last cabin on her route, she was alternatively in a felicitous mood or depressingly maudlin. One elderly and kindly matron returned to her cabin and discovered the cabin girl in the latter mood. In fact, she was sitting on the sofa, sobbing. The matron sat down beside her, patted her solicitously on the knee, and inquired, "What ever is the matter, my dear?"

Between sobs, the cabin girl blurted out, "I'm not getting any, and all the other girls are."

Not understanding the phrase *getting any*, the matron asked, "What is it that your are not getting, my child?" The cabin girl stared at her open mouthed.

Assuming that she was referring to appropriate equipment or overtime pay, the matron said, "I'm sure that if I speak to Mrs. Horton, she can get you some." At that, the cabin girl stumbled out the door shaking her head.

A wrangler and a cabin girl were once smitten by either Cupid's arrow or a case of Sheridan Export and decided on the spur of the moment to get married. Realizing that the nearest justice of the peace was thirteen miles away and probably unavailable on a weekend, they suddenly remembered that a Supreme Court Justice was in residence on the ranch and probably outranked a justice of the peace. They eagerly sought him out and found Justice Stewart ambling down the path. When asked if he would marry them, the Justice immediately sensed that this would not be a great idea. Hoping to transfer the problem, he noticed Admiral Greer strolling by and called him over. The Justice said, "This fine young couple has asked me to marry them, but I'm afraid I lack jurisdiction. As an admiral, you have authority on board ship which by definition, means you are surrounded by water. Therefore, if you were to stand in the middle of the creek, perhaps you could officiate."

As the admiral found the idea equally unappealing, they both launched into a long, involved and highly intellectual discussion of landed and marine jurisprudence. Eventually, the young couple became bored with the entire process and were persuaded that they could wait until the next week. In the meantime, Cupid moved on to greener pastures, and the marriage never took place.

The staff was, and is, a professional, cheerful, fun loving group that does its best to make the dude welcome. The occasional lapses or misadventures are easily surpassed by those they serve—the guests.

THE GUESTS

Shortly after dinner, the lady arrived at the corral and mounted Tarbaby, a short, fat, sleek, and very dark horse and then headed off to the Speedway, her favorite evening ride. She was accompanied by two children and a young woman. Everyone has a favorite ride, and the Speedway in the evening seemed to offer her three defining characteristics of the West—vastness of space, variety of landscape, and a sense of the aridity that reinforces the precious nature of water.

Like many trips to vantage points with beautiful views, the ascent can be a dull grind unless you permit your senses to be immersed in the surroundings. Antelope will spring out of nowhere and disappear with their cotton puffed hindquarters flashing. Mule deer will give you a curious gaze and then move away in spring-like leaps as if each leg were a pogo stick.

Everything seems to slow down and become more distinct as the sun sets. Cooler currents of air can be felt and the sounds of birds and insects are magnified. A merlin may skim above the ground and dive down a hill in a last hunting foray of the day. A common night hawk in courtship display may dive to the ground feathering its wings to produce a fluttering sound and pulling up at the last possible second.

From the Speedway, which is an open strip of land nestled in the foothills of the Big Horns, you get a magnificent view of the plains. The heat of midday makes the plains hazy and blurs the soft colors typical of an arid region. Sunrise and sunset cast shadows that show the plains distinctly in relief, adding dimension and sharpening colors.

On reaching the Speedway, the lady suggested to her companions that they dismount to enjoy the mystical moment of sunset. As she dismounted, the saddle slid toward her and when she reached the ground, the saddle continued on until it was under the belly of the horse. Tightening a cinch is more art than science. The corral quite naturally prefers a looser cinch so the horse does not develop sores. The guest generally ranks personal safety higher on the scale than the horse's comfort and prefers a tighter cinch. On rare occasions, a guest or a wrangler has ended up walking home.

On this occasion, Tarbaby, a normally phlegmatic horse, contemplated the saddle in an abnormal position and then did what a horse typically does when confused and scared. He went berserk. With eyes wide opened and ears flattened in terror or rage, he went into a bucking routine that no one would have thought possible from such a short, fat horse.

Then the impossible happened. The odds were similar to a monkey randomly striking the keys of a typewriter and producing Shakespeare's *Othello* or a poker player drawing a royal flush five times in a row. Tarbaby somehow kicked each hind hoof through the stirrups of the saddle and, finding himself hobbled in midair, went down in a crash. A brief struggle seemed to prove that further efforts would be futile, and he quieted down.

While the young woman sat on his neck patting him and uttering soothing words, the lady attempted to extricate Tarbaby's hooves from the stirrups. Unfortunately, the stirrups were like well-fitted shackles. Although everyone gave it a try, no one could slip a stirrup off a hoof. Finally, the young woman rode back to the corral to seek assistance. When she told the story to Dean, he looked at her skeptically and said, "You've got to be kidding."

Dean was persuaded to return with her and bring a tool kit. When he reached the Speedway, he found the lady sitting on Tarbaby's neck muttering soothing sounds. Each hoof was still jammed through a stirrup. Dean's first reaction was to burst into laughter and say, "I'll be damned." He then tried to pull each hoof from the stirrups but finally gave up. Out came the tool kit, and with pliers and wrenches, he dismantled each stirrup and extricated the totally perplexed Tarbaby. He then reassembled the stirrups and was about to return to the ranch when he stopped, slapped his forehead, and said, "Nuts. I forgot to borrow a camera." When asked why he wanted a camera, Dean replied, "They'll never believe this down at the Elk's Club."

Yes, this really happened. The lady was my mother. Unfortunately, I was backpacking at the time in the Bear Tooth Mountains of Montana, but I have it on excellent authority.

Most stories about ranch guests revolve around incidents with horses, which is not surprising as the guests ride twice a day, in the morning and evening. The safety record is astonishingly good because horses are matched to abilities; professional assistance, supervision, and training are readily available; and the horses are part of a ranch-owned herd so they are used to each other. Nevertheless, a few cynics including myself feel that when you are on top of twelve-hundred pounds of latent stupidity, occasional incidents are a reasonable statistical bet. Over the past fifty plus years, I have been unseated several times with no damage to myself other than ego. I am not a superstitious man, but I find myself knocking on wood as I process these words. However, I would guess that I am safer on a horse than driving the interstate downtown. I believe that intellectually but not viscerally.

One of the few injuries involving a guest happened to an excellent, experienced, and cautious rider. The man was my uncle Truman. He was thrown on the morning ride and broke his wrist. Although the break was perfectly clean, there was cause to worry as Truman was a principal violinist in the Omaha Symphony Orchestra.

Now as it turns out, if you are ever in an accident or ill, Buffalo, Wyoming, is the place to be. The local hospital is excellent and staffed by highly competent doctors and nurses who also know how to be good human beings without trying. A broken bone is common fare, but they are superb on anything from ulcers to heart attacks. If I am ever struck with some affliction, I pray it happens near Buffalo, Wyoming, as opposed to say, New York City.

As Truman was alone at the ranch that summer, one of the staff drove him to the hospital where the bone was set. His wrist ultimately healed perfectly, and some of his uncharitable colleagues said that he played better once his wrist had been fixed by the Buffalo surgeons. Truman was rather loopy from anesthetics and pain killers when a ranch car returned him to his cabin that afternoon. Shortly after his arrival, Hank stopped by to see how he was doing. Truman fixed her a drink, and they chatted amiably for about an hour.

Finally, Hank said she had to accomplish some chores before the dinner bell rang and took her leave. Truman said that he would freshen up and join her shortly for dinner. He stepped into the bathroom and noticed for the first time as he passed before a mirror that he was not wearing a stitch of clothing. Not wanting to embarrass Truman, Hank had remained silent on the theory that he would discover his state before real damage was done.

The only broken bone from a horse incident that I have ever witnessed occurred on the return ride from a pack trip to Willow Park. We were picking our way slowly through the rocks of the North Fork Canyon, and one horse was giving its rider a particularly difficult time. The horse "smelled the barn" and was so eager to get home that he had to be constantly reined in as he was dancing and stumbling on the rocks and tossing his head. In a fit of perfectly understandable frustration, the rider gave the horse a karate chop between the ears. I initially felt that this was a perfectly rational way of dealing with an animal of a horse's intelligence until I remembered that horses have very thick skulls. This horse was totally indifferent to the blow, but the rider broke his hand on the animal's head. Such an injury is difficult to explain at cocktail parties. Since then, I have noticed that action figures in Kung Fu movies break boards and bricks with their bare hands, but they never take on a horse.

Incidents with horses, particularly when no one is hurt, usually provoke a good deal of mirth and tongue wagging. Consequently, unseated riders often go to great lengths to rationalize their misfortune, for example, "When the mountain lion shot across the trail in front of my horse, he started bucking furiously." One woman of exceptional intelligence, candor, and integrity happened to mention to Hank over cocktails that she had fallen off her horse that morning. Hank looked at her in total amazement and said, "You are the first person in the state of Wyoming to fall off a horse. Everyone else was thrown."

Although interesting people always have fun and are never bored, the manner in which they amuse themselves has changed over the years. During the era of gas rationing during and immediately after World War II, families would arrive by train with fully packed steamer trunks and spend the entire summer at the ranch. Husbands would drift back to offices but still seemed to have much more leisure time than in today's restructured and downsized environment.

With plenty of time on their hands, guests would engage in heavily organized activities such as elaborate costume parties, rodeos, or serious baseball games against the dudes at Eaton's or Paradise. There would be gourmet picnics with truite bleue and other delicacies. Dressed outrageously in Indian wear borrowed from Hank's collection, the corral might surprise a picnicking group by swooping out of the bushes with fearsome war cries followed by demands for a share of the food and libations.

As vacations shortened and time became more pressing, activities became less organized and fun just seemed to happen spontaneously. The following would be a fairly typical example.

Following the evening ride, two couples strolled back to their cabins enjoying the deepening nightfall with stars appearing at astonishing proximity given the clarity of the atmosphere. Wanting to prolong the moment if ever so slightly, one couple suggested a short night cap. After the obligatory protestations of, "Well only one, because we have a long day tomorrow," the bowl was filled with ice and the bottles uncorked. After several days without television, radio, telephones, and beepers, the sound of the stream and the rustle of the breeze through the trees return the spirit to its natural state. One even begins to realize that a long day tomorrow cannot exceed its normally apportioned twenty-four hours.

As the levels in the glasses declined to their unfortunate conclusions and the discussion became more scintillating, one dry guest suggested that the evening was young and perhaps a "halfsy" was in order. This suggestion was greeted by weak demurrals, but conversation ceased with the sound of boots scraping across the boarded porch, a knock at the screen door, and a raspy voice saying, "Anybody home?" When you are standing on a dark porch peering through a screen door into a fully lighted living room, it does not take long to conclude that someone is home. Consequently, Dean strode in and was unnecessarily waved over to the bar where he poured himself a bourbon and branch and effectively resolved the issue of whether or not another drink was in order.

By the age of thirty, Dean had packed more into life than most nonagenarians would ever experience. He had been around many proverbial blocks and burned quite a few candles at both ends, so the prospect of quiet conversation over night caps raised the possibility of hearing more stories and filling in some of the chinks in a fascinating life. One of the guests decided he wanted to hear about the time Dean and Clark Gable went on a bender, so the conversation was herded in that direction.

One guest asked innocently, "Didn't you once work at a dude ranch in Arizona where all the Hollywood crowd went?"

Dean thought for a moment and replied, "That was the Silver Bell Ranch about five miles out of Tucson. Of course, Tucson grew right over it, so it doesn't exist now. I was out riding once with three other guys. One was Groucho Marx.

Guest: "You mean Groucho went there?"

Dean: "Oh sure. He rode a horse pretty good. Clark Gable was a guest. He was a good guy. We had more damn fun. The other was a nice fellow from Denver who made shoes for crippled people. Had a big factory in Colorado. Anyway, we were riding along when we saw a herd of about thirty pigs."

Guest: "I think they are called peccaries."

Second guest: "No, I think they're called javelinas."

The third guest realized that the train of thought was at a dangerous crossing and perilously close to derailment. The animal in question is actually a collared peccary, but it is more normally called a javelina in Arizona. In the southern part of its range, it is called a collared peccary to distinguish it from the white-lipped and Chacoan peccaries, which are native to Central and South America. The animal belongs to the zoological family Tayassuidae, while the true pig belongs to the family Suidae. Realizing that if the conversation veered toward precision in the classification of the Artiodactyla order, they would never get to Clark Gable, the third guest blurted out, "So what happened with the pigs?"

Dean: "Well, they all took off in a hurry when we rode up, but one little pig was left behind. Just a small guy. I said I would catch it, but Gable said he wasn't going to get off his horse. The other thirty might come back. He told me I better stay on my horse, but I got off and ran the pig down and put my hat over it. They have sharp teeth, you know. I finally grabbed it by the back of the neck. The guy from Denver had an old denim shirt, so he tore it up into strips and we tied the pig up. I had quite a time riding with the pig back to the ranch because I was riding this horse that wasn't too well broke yet. Anyway, I put the pig in my horse trailer but he wouldn't eat anything. We gave him manzanita roots and carrots, but all he would do is bite a stick. I didn't know what to do. I was about to turn him loose when Clark said let's go to Mexico. I had a half day off, and we returned two days later.

Guest: "Where did you guys go?"

Dean: "Well, we thought we might go to Sonora, but we ended up in a bar in Nogales. It was kind of a cavern. It used to be a prison and still had chains on the wall and a real long bar. Clark had on a western hat with his shirt collar turned up so no one would recognize him. We decided to have something to eat, so Clark ordered a bottle of scotch and I got a bottle of bourbon. They brought us plates of some kind of food. Anyway, pretty soon some little girls came over and asked for autographs, so we got out of there.

Guest: "Where did you go?"

Dean: "Well, let's see. We went to Bud and Norma's. They had a dude from the ranch that built a guest house on their property. She was the one that sent me to art school."

The guests looked at each other and silently concurred that they would pursue the art school story later. The possibilities of that one looked promising, but they wanted to hear the end of the current story.

Dean: "Well Clark drank a whole bottle of scotch and slept in the guest house. The next day we went on to see the one-legged rancher (another story) and stayed a day. I only had a half day off, and we were up to two days now. We were driving Clark's Cadillac convertible because he didn't want to ride in my truck. They had the highway patrol and everyone out looking for us because they thought we had been in a wreck. We were pretty tired and didn't feel so great when we got back to the ranch."

Guest: "What happened to the pig?"

Dean: "Oh, he died."

Guest: "I'm sorry I asked."

Dean: "Well, you know, I put him in the barn, in a little manger, with a lot of stuff to eat, but when I came back, he was dead."

Guest: "Why did you keep him?"

Dean: "Well, if you get them young enough, they make great watch dogs. If someone comes, they grunt and make the damnedest noise."

Before the guests could fill in the details of the Dean and Clark toot, more boots thumped across the porch and several other people dropped in followed in short order by another handful. Because the cabins are close to the creek, the sound of rustling water muffles most noise. However, the banging of a screen door, boots on the porch, the creaking lid of the icebox, and the chip, chip, chipping of pick on ice are pretty good indications that something is going on.

People materialized out of nowhere and further storytelling was impossible. The gathering mushroomed into a full-scale party and pretty soon, people were coaxing Dean into doing his one-armed fiddler routine.

Although Dean's acting talents were perhaps untutored, he did have a sense of timing that indicated that the one-armed fiddler should only be performed after the audience's sense of propriety had been considerably weakened by alcohol. He and an accomplice disappeared into another room where Dean costumed himself in a jacket with a pinned-up sleeve to indicate the lack of an arm. The real arm was pushed down into his blue jeans with his hand poised at the unbuttoned fly. The two reappeared with the accomplice playing the role of a conductor about to perform a violin concerto with Dean being the guest violinist. The conductor and Dean bowed to their audience, and the conductor then handed Dean a prop to indicate a violin, which Dean took in his left hand. The conductor then handed Dean a bow, and Dean went through a few contortions trying to figure out how to grasp it. Finally, in a fit of exasperation, he reached a gnarled finger through his fly and hooked the bow.

Admittedly, most readers in the cold light of day will find this performance somewhat crude. However, when delivered at the appropriate time and under the right circumstances, audiences invariably find it hilarious, and this time was no exception.

After several more trips to the porch to replenish the ice bowl, the guitars came out, and launched into a few railroad songs such as "The Wreck of the Ole 97" and "Wabash Cannonball" for which Dean did his deep-throated imitation of a train whistle. Dean had no ability to carry a tune, but his gravely voice was always eagerly awaited in the part of the chorus of "Frankie and Johnny" when he rasped out, "But he done her wrong." The inevitable request for "Cool Water" prodded the musicians into a valiant effort that produced a version that would have the Sons of the Pioneers spinning in their graves or wherever they happened to be at the time. When they reached the line that went "Old Dan and I with throats burned dry, and souls that cry for water," the screen door would swing open and Dean would crawl into the room gasping and pleading for water. He normally got more bourbon.

A little later, Jimmy and his wife Pearl arrived. Jimmy was a friend of Dean's and had worked at the corral on several occasions. Pearl spoke in a few well-chosen monosyllabic phrases such as "Scotch, please" while Jimmy pulled out his guitar and tuned to the current group of musicians. He would listen to a few songs and then say, "O.K. This is the way us real cowboys do it." Jimmy happened to be a welder at the time, but once a cowboy, always a cowboy. His best contribution was a song about a girl from the Yukon that had a chorus that went, "Ouka Ouka Moushka/ That means that I love you/And if you'll be my baby/ I'll Ouka Ouka Moushka you." Everyone belted out the chorus with gusto, and a few even acted out their interpretations of "I'll Ouka Ouka Moushka you" with astonishing effect.

As the hour was quite late, the only addition to the party was a liver-colored springer spaniel whose owners had left hours ago. The dog, who was named Henry after Hank Horton, enjoyed people, attention, and the occasional drink that might be left within lapping distance. He had a fine time. After a number of limericks that would make a mule skinner blush sung between choruses concerning the absence of chili in China, everyone drifted off to their own cabins. A lady nicely volunteered to walk Henry home to Branding Iron, which has a small bridge across a minute brook. Henry, who had navigated this bridge by day and night for years, fell off and had to be rescued by the lady. As both were soaked, the lady pushed Henry through the front door and returned to her cabin to sleep through the few remaining hours of the night.

The next morning, one of Henry's owners padded into the extra room where the spaniel slept rather grandly on the bed. The lady thought it quite odd that Henry was still asleep as he normally was the first up demanding some breakfast and then a walk. When she shook Henry, she noticed that his coat was all wet and matted. Henry slowly lifted his head and raised his lids to reveal bloodshot eyes. And then the unthinkable happened: He growled at his loving mistress.

Other routines were slightly askew that morning. Much of the dining room seemed in unnatural quiet. An ashen-faced doctor entered the room and stopped by the table seating the occupants of the cabin where the party had occurred. He said to them, "I meant to thank you last night, but I couldn't pronounce it."

Another man sat down groaning next to the doctor and began describing his symptoms in clinical detail. The doctor listened in silence and then said gravely, "I diagnose your ailment as a case of the whiskey shits. That will be $50 please." The couple who originally said they would have just one night cap did, indeed, have a long day.

The random, unplanned convergence of guests, staff, and other catalysts often produces memorable events and the stuff of countless stories. I am no longer surprised that when one guest meets another from a different part of the country, they often have mutual acquaintances or colleagues. A couple once went to Willow Park, and the packer was making dinner when along came a bedraggled pair of backpackers. They had been in the mountains about a week and brought little food as they intended to live off the land. Living off the land is always a mistake, since trout are unpredictable and have very low caloric value, particularly for young men with active metabolisms. Taking pity on the pair, the packer gave them some bread, a large can of jam, and other items to impart high energy quickly. After wolfing down prodigious amounts of food in record time, they sat back and chatted gratefully with the packer and the guests. After the initial inquiries about the location of homes, it turned out that all five lived with twenty-five miles of each other in the Washington, D.C., area.

Many guests have led fascinating lives or are truly unforgettable characters. In the latter category was an octogenarian who sported a real or imagined military title. He loved to fish, and each year he would arrive at the ranch in a rented car that he would use to drive to stream side. Barely able to peer through the steering wheel, he would careen down the gravel roads, veer off at a promising spot, and scrape over rocks and debris until the vehicle came to a protesting stop. Returning to the ranch with a fish, he would clean it in his toilet, which naturally gave the cabin girls the vapors. Quite frequently, he would become stuck by the side of the stream and wait patiently until a ranch vehicle came to winch him out. Each year, he had to rotate car rental agencies because of the shape of the vehicle he returned. He also had the bizarre habit of propositioning any younger woman he met. No one was sure if this was a manifestation of capability or the unconscious remembrance of times past. Each year, he became a little more decrepit, but to the end, he always kept trying.

Guests come from everywhere and always manage to bring a little something to the residents of Wyoming and vice versa. An Italian family enjoyed the western experience for many years. One summer when the boys became quite intrigued with tales of cowboys and Indians, the father gave them bows and arrows. They had read about Native Americans fishing with arrows, so they thought they would try it at Lake DeSmet. While the boys scampered along the shore shooting arrows harmlessly into the water, the father watched them from a comfortable vantage point. Pretty soon, a cowboy appeared and engaged the father in idle conversation. The cowboy noticed that the father spoke with an accent and the boys were dark skinned, albeit from many days spent in the full summer sun. Assuming they were Native Americans, the cowboy inquired, "What tribe are you?"

Pre-Etruscan groups in what is now present-day Italy were often referred to as tribes, and one of the more prominent was the Veneti. Thinking this was one well-informed cowboy, the father said, "We're Venetian."

The cowboy thought for a moment and said, "Venetian? I'll be damned. Sioux, Crow, Cheyenne, Arapaho, Shoshone, Ute. I thought I knew 'em all."

With interesting people, the dining room conversations cover an amazing variety of topics, and from her vantage point, Hank unintentionally overheard a number of them. Tom Baird, who was a college professor, art historian, and novelist, once delivered an impromptu lecture on the Donner Party for which the pass was named. The group had become trapped in a sudden snowstorm, and the survivors resorted to cannibalism to stay alive. Although the facts are somewhat murky, Tom maintained that it was morally acceptable to eat the departed in order to survive. In fact, he had no qualms about his remains being so consumed in similar circumstances. One of his listeners looked at him dubiously and said, "Have you ever tried to clean a college professor?"

In an amazing tribute, the guests arranged a surprise seventy-fifth birthday party for Hank that was held at the Brown Palace in Denver. This was a costume party and people poured in from all over the country. Jack Horton was dressed in furry sheepskin chaps and a World War I campaign hat while Dean wore full plains Native American regalia, all of which was original and came from the ranch. There were German officers with dueling scars, gypsies, clowns, Beefeaters, and you name it. My aunt and uncle, Sis and Truman, went as the servants from *Upstairs Downstairs*. Looking courtly in a tuxedo, Truman was treated with great deference by the hotel staff. When Sis, who was dressed in a maid's uniform, went to the gift table to rearrange some ribbons, she was reprimanded and admonished by a hotel staffer to stay away from the guests' gift table. She sighed and said, "Well, I guess that answers that. Clothes do make the woman."

One of my favorite memories of that evening came when Hank was confronted with cutting a huge birthday cake with a minuscule knife. A guest magnificently attired in a confederate officer's uniform gallantly rose to her assistance, drew his saber over his head, and lurched toward the cake. The evening was filled with such moments, and everyone had a great time. The July people also met the August crowd who had always referred to the July guests as "those people who dulled the ice picks."

The participants in that magic evening are moved even today by the camaraderie, humor, and fun that bound people together. It was, indeed, recognition by interesting people of a wonderful and fascinating woman who provided friendship, laughter, and the western experience.

AFTERWORD

The workmen had finally finished the task of cleaning out all of the dead wood, overgrown brush, and other debris in the area of the swimming pool. Sweaty and tired, they presented their bill at the adjacent office where Hank and Margi Schroth were going about their administrative chores. Turning to Margi, Hank asked, "Do you want to write the check or should I?"

Margi answered, "I'll be happy to take care of it."

Scowling, Hank retorted, "Well, you'd better write it in red ink because that's what will be flowing around here."

The burdens of running a large operation were beginning to tell on Hank. The rigors and extremes of the Wyoming climate demanded constant repairs and reinvestment in the physical plant. As Hank could be a little tight on capital investment, these expenditures pained her almost as much as the administrative problems of getting the actual work done. And then Jack, who provided the continuity for the ranch, died in 1981. Hank was seventy-seven at the time, and the ranch was truly at a crossroads.

Fortunately, Jack had met Margi on a trip to Billings where she was working in public relations at a local hospital. Formerly from New York State, she had a degree in English from Briarcliff and had been a magazine writer in the Big Apple. After working on several ranches in Montana, she moved to Billings, but after meeting Jack, she joined the HF Bar staff. She and Jack were to be married, but that dream ended with Jack's leukemia.

When Margi inherited Jack's share of the ranch, she and Hank ran it together until Hank's death in 1989. A big-hearted woman with boundless energy and vision, Margi restored the physical plant in ways that were both visible and transparent to the guests. Plumbing, wiring, and kitchen equipment were not particularly noticeable but were essential nevertheless. Bridges were repaired, fences mended, and trails cleared. There was an old, leaky swimming pool with a coal-fired heater that raised the water temperature from bone chilling mountain stream temperature to borderline unbearable. It was replaced and enhanced with landscaping and other improvements.

Along with the physical changes, the ranch once again became a place of wonder for children so that generations of family could enjoy the western experience together. In many ways, HF Bar has gone full circle returning to the atmosphere Skipper and Hank established in their youth. Many of the same traditions are still observed.

Some things never change. The cry of a prairie falcon over Castle Rock, the brilliant blue and perfume of lupine in the early spring, the bold relief of the plains at sunset. While new generations forge their own recollections, we say to Skipper, Hank, Dean, Jack, and Trudy, thanks for the memories. Paraphrasing from Dean's favorite song*:

Have we told you lately that we love you
Could we tell you once again somehow
Have we said how much those times have meant to us
Well, good friends, we're telling you now.*

*Scott Wiseman, "Have I Told You Lately That I Love You," 1945.

Edgar M. Morsman, Jr., a consultant to the banking industry and a retired executive, is the author of several books about commercial lending. Ed is also an experienced wilderness traveler and fly fisherman who has backpacked extensively in the Big Horns as well as other ranges in Wyoming and Montana. He and his wife, Beth, live in Deephaven, Minnesota.

Ed would like to acknowledge Beth Brodovsky, graphic artist, and Shelley Wilks Geehr, editor, for their contributions to this publication.

Original pen and ink by Dean Thomas.